As the World Turns

The History of Proving the Earth Rotates

T0338657

As the World Turns

The History of Proving the Earth Rotates

Peter Kosso

Northern Arizona University, USA

World Scientific

NEW JERSEY · LONDON · SINGAPORE · BEIJING · SHANGHAI · HONG KONG · TAIPEI · CHENNAI · TOKYO

Published by

World Scientific Publishing Europe Ltd.

57 Shelton Street, Covent Garden, London WC2H 9HE

Head office: 5 Toh Tuck Link, Singapore 596224

USA office: 27 Warren Street, Suite 401-402, Hackensack, NJ 07601

Library of Congress Cataloging-in-Publication Data

Names: Kosso, Peter, author.

Title: As the world turns : the history of proving the Earth rotates /
    Peter Kosso, Northern Arizona University, USA.

Description: Hackensack, NJ : World Scientific, [2020] |
    Includes bibliographical references and index.

Identifiers: LCCN 2019047676 | ISBN 9781786348173 (hardback)

Subjects: LCSH: Earth (Planet)--Rotation.

Classification: LCC QB633 .K67 2020 | DDC 525/.35--dc23

LC record available at https://lccn.loc.gov/2019047676

British Library Cataloguing-in-Publication Data

A catalogue record for this book is available from the British Library.

For any available supplementary material, please visit
https://www.worldscientific.com/worldscibooks/10.1142/Q0242#t=suppl

Typeset by Diacritech Technologies Pvt. Ltd.
Chennai - 600106, India

Printed in Singapore

Preface

Two things you might remember from your high-school science class. The first you knew already, but that didn't stop the teacher from celebrating the proud victory of modern science over ancient and medieval dogma: The Earth moves. It both revolves around the Sun and rotates on its own axis. The other thing you might have known as well, maybe with less clarity, because it's a bit more abstract: All motion is relative. It doesn't make sense to say that something does or doesn't move without identifying, at least implicitly, the reference relative to which it moves, or doesn't. This may be as far as you got into the theory of relativity, but there you have it, the reason it's called relativity.

Don't feel bad if you didn't notice that these two fundamental ideas are incompatible. We usually learn them and talk about them in widely separated sections of the book. The motion of the Earth is near the beginning, when modern physics is put together by Copernicus, Galileo, Newton, and their contemporaries. It's been obvious ever since. The relativity of motion is more subtle, and though it really starts with Galileo, we read about it late in the development of physics and associate the idea with Einstein.

Bring the two together and you should feel the tension. Saying that the Earth moves is saying that Copernicus was right. The Catholic church and the legacy of Aristotelian cosmology were wrong; they had the Earth motionless at the center of the world. Galileo was not brought before the Inquisition on suspicion of defending the idea that the Earth moves *relative* to the stars. The challenge to scripture and the claim of the Copernican model were in saying that the Earth moves, *period*. A strict application of the relativity of motion makes that claim meaningless. It says that any object can be the stationary reference of motion. That includes both the Earth and the Sun. The disagreement between the two world systems is moot.

The way to sort this out is by following the evidence. This is science, after all. As individuals there's nothing we can see or feel that unambiguously shows the rotation of the Earth. The phenomenon is not directly observable in that sense. Maybe it is in some way indirectly observable. Galileo famously enhanced the quality of astronomical observation by using a telescope, but no telescopic image showed the Earth moving. His case was not simply a matter of finding the right object or objects to view and saying, Look! The Earth moves!

The challenge was in figuring out that something is moving while standing on that something. We know the difficulty from the experience on a smoothly moving train, just leaving the station. Are we moving or is that train next to us moving? By the principle of relativity, though, it makes no difference. We are each simply moving with respect to the other.

Waiting for evidence on the rotation of the Earth until someone could leave the planet and look back would put Galileo in a vulnerable position. The first man in space was Yuri Gagarin, in 1961. And of course, he was in orbit around the planet, as are the astronauts in the international space station. The orbital motion complicates the report of the Earth's rotation. You wouldn't claim to be seeing the Earth move as you watch the ground pass under an airplane, since you have to consider your own motion. Observations from the space station are going to have to at least compensate for the motion of the station itself. It's not just a matter of pointing down and saying, Look! The Earth moves!

So we'll follow the evidence on the rotation of the Earth, from what the ancient Greeks had to work with to the best we have now. It will be necessary to consider things in context, to view the data in light of the scientific understanding at the time. That's how science is done, using the best theoretical information to make sense of the best observations. The historical trail will facilitate a pivotal question.

When and how did it become reasonable to believe that the Earth rotates? Put it the other way around, if you like. When and why did it become unreasonable to maintain that the Earth stands still?

What follows is the answer—expanded version—you might have heard if you had asked your science teacher this question: I understand that the Earth rotates, but how do we know that it rotates?

There is a shorter version of the answer available at the end of the book. The Timeline of Important Events and Ideas lists the evidence that the Earth rotates, in the scientific context it occurred. The main events, the sorts of things most science teachers will mention, are in bold.

Contents

About the Author

Peter Kosso is a philosopher of science. He taught physics at Montana State University, and taught philosophy first at Northwestern University, and then at Northern Arizona University. He is the author of *Reading the Book of Nature*, *Appearance and Reality*, *Knowing the Past*, and *What Goes Up...gravity and scientific method*, as well as numerous articles on relativity, quantum mechanics, astronomy, and scientific method.

CHAPTER 1
On Uncertainty

The crucial thing is being able to move the earth without causing a thousand inconveniences.
—**Galileo**

In the time that Galileo spent in house arrest, between his conviction on June 22, 1633, being "vehemently suspected of heresy," and his death on January 8, 1642, the Earth turned on its axis 3,122 times. He was confined to his home in Arcetri, just outside Florence, for what turned out to be almost 8 years, that is, 3,122 days. That's the number of full rotations of the planet, using the Sun as the reference. If we use any other star as the reference, Spica, say, or even the whole field of stars, then the number is different. In the same time, the Earth turned around 3,130 times.

There's nothing complicated or mysterious here. It's simply the number of times the Sun has risen or set, or the number of times Spica has risen or set, between the two events of Galileo's conviction and death. That's from our own perspective. Another way to look at it is to consider the view of the Earth from the Sun or from Spica. From the Sun, we would see the Earth turn fully 3,122 times around. From Spica it would be 3,130. Like any determination of motion, it depends on the point of reference. Astronomers keep track of this by distinguishing between a solar day, one complete rotation of the Earth with respect to the Sun, and a sidereal day, one complete rotation of the Earth with respect to the stars.

So, how many times did the world turn fully around, really?

Galileo's accusers, of course, would say the right answer is exactly zero; the Earth didn't rotate at all, since the Earth doesn't move at all. And this is true, from any point of reference on the Earth itself. If there is a reference frame in which the number of rotations is 3,130, and another in which it's 3,122, there is a system in which the number is zero. This is the reference frame attached to the object itself. We can ask if one of these reference frames is in some way distinguished as being the one that's right, while all the others are distortions of the truth. If there is, which one is it? Or better, what are the criteria to determine the proper reference of rotation? Knowing the rules, we can apply them to other cases. And, the more we know about the basics, the less we have to rely on authority about the results.

What difference does it make, how many times the Earth rotates, how fast it rotates, or whether it moves at all?

It made a very big difference to the Catholic church. The issue was important enough to threaten Galileo with torture and lock him up for life as punishment for what they regarded as advocating the new world system, the Copernican model of the cosmos. Peer review was different in the time of Inquisition.

The rotation of the Earth is a big deal for most junior high school science teachers, too. They're pretty clear about it, seemingly just as certain that the Earth does move as the Catholic church in 1633 was that it does not. And if you get it wrong on the quiz there will be consequences, although less severe than suffered by Galileo.

The Catholic church had scientific allies in its claim that the Earth stands still. The school of Aristotelians had been the authority for both physics and astronomy for two millennia, and like the Catholic church it was not accommodating of uncertainty or open to challenge on the stability of the Earth. Aristotle did science the way mathematicians do proofs, with certainty. It's this methodological standard, as much as the conclusion about motion that separates Galileo from the Greek. In his defense against the charges brought by the Inquisition, Galileo claimed to be demonstrating only that the matter of the Earth's motion was uncertain. He was not, he said during the trial, advocating for one side or the other. In his fourth and final deposition to the court, he explained

> In regard to my writing of the Dialogue already published, I did not do so because I held Copernicus's opinion to be true. Instead, deeming only to be doing a beneficial service, I explained the physical and astronomical reasons that can be advanced for one side and for the other; I tried to show that none of these, neither those in favor of this opinion or that, had the strength of a conclusive proof and that therefore to proceed with certainty one had to resort to the determination of more subtle doctrines, as one can see in many places in the Dialogue.

He was referring to the publication that got him in trouble with the Catholics, the *Dialogue Concerning the Two Chief World Systems*. The old Aristotelian world system had the Earth stationary at the center of things. The new Copernican system put the Earth in motion, revolving around the Sun and spinning around its own axis. A dialogue is an unusual way to present scientific ideas. It is now and it was then. But it seems an appropriate presentation of competing claims, their strengths and weaknesses, the reasons for and against, and the resulting uncertainty in deciding which is correct. This was Galileo's defense.

Not only had Galileo given voice to both sides of the issue, he had them speak in the vernacular, in Italian rather than Latin. This made the debate accessible to more than the scholars who had been trained in the tradition of the old world system. By presenting both sides, it might seem as if Galileo was turning science into a democracy. He wasn't, and like his exemplars of philosophical dialogue, Socrates and Plato, he would have been appalled at the suggestion that important questions could be answered by popular vote. Galileo knew there was a right answer to the question of whether the Earth moves, and he was convinced he knew what it was. The point of the *Dialogue* was not for the readers to simply decide for themselves, but to consider the evidence and arguments and then decide for the new world system. Uncertainty doesn't mean there is no good reason at all—I'm not absolutely sure the water from my kitchen faucet is safe to drink, but I am reasonably justified in my confidence that it is—and Galileo clearly presented the evidence that favored the new world system.

The *Dialogue* had a scientific point to make; that doesn't mean it wasn't also about the uncertainty of the situation. But the Inquisition didn't buy Galileo's defense. They saw only the advocacy of the new world system, and it's easy to see why. The debate between the two world systems was neither fair nor balanced. The Copernican system with the Earth in motion ends up a clear if undeclared winner in the *Dialogue*. The stationary Earth never had a chance. There are three characters in conversation throughout the book, an advocate for each of the two world systems and a more or less neutral moderator who speaks when clarification is needed. The names of the speakers are telling. The Aristotelian is Simplicio, possibly after a real sixth-century philosopher Simplicius, but certainly for the implication in the meaning of the name. The Copernican is Salviati, as in salvation. The third party, a supposedly open-minded layman, is Sagredo, in fact a friend of Galileo's.

The church hierarchy saw no uncertainty, since its conviction—allowing for two meanings of the term—was based on both scripture and basic observation, with more authority to the former. In the formal condemnation of Galileo, it was made clear that to claim the Sun is the center of the world and the Earth is in motion is "contrary to the senses and Holy Scripture."

Modern science textbooks are pretty sure in their presentation of the Copernican model. There is no dialogue that presents both sides. In the authoritative style of textbooks, there is no invitation to doubt. The authority is not scriptural, although the textbook itself often takes on that status. The conclusion is based on evidence, but not the basic observations cited by Galileo's accusers and Aristotelians. It must be more sophisticated evidence than simply standing, two feet on solid ground, and watching the Sun rise. And this sets up our task, to follow the connection between what is observed and the conclusion that the Earth rotates.

We will do this historically, following the evidence from well before Copernicus up to our own time, asking when did it become reasonable to believe that the Earth rotates. It might help to rephrase the question. When did it become unreasonable to deny that the Earth spins on its axis?

Science textbooks usually have little interest in the history of the subject. The only reason to describe the mistaken attempts of the past would be as a foil to clarify the correct theories we have now and maybe to gloat a little. This attitude is companion to the overall confidence that what we have now is correct, no doubt about it. There is sometimes a short description of scientific method in the first chapter of a textbook, including the disclaimer that it is impossible to prove a scientific theory. This cautious skepticism has a way of disappearing in the rest of the book, either implicitly in the assured language about the scientific content, or explicitly in describing crucial experiments that prove this or that. Maybe the tone of certainty is justified in a simple thing like the rotation of the Earth. Maybe by now this is something we can just observe and know for sure, nothing circumstantial about the evidence. We'll have to see.

Acknowledging uncertainty in science was one of Galileo's most significant contributions. He is often described as a pivotal player in the scientific revolution and the origins of modern scientific method. Accounts of his role can be simplistic, suggesting he was the first to emphasize the importance of empirical data, or the first to do an experiment, or simply put, the first to do anything we would call science. He wasn't. It's not that he didn't observe and experiment. He did these and he did them well, but so did a number of people, a number of scientists, before him. Some of them did these things responsibly and yet concluded that the Earth does not move. Furthermore, Galileo sometimes ignored the evidence, and even explicitly argued for the occasional need to ignore the evidence. Some of his most important results were derived with no evidence whatsoever, by ingenious application of logic.

It's Galileo's deft combination of evidence and reason that is his mark on the scientific revolution and his contribution to the method. That, and explicitly pointing out the resulting uncertainty. The emphasis on the empirical does not put science on unassailable foundations. As we'll see, evidence and observation are always clouded by a need for interpretation and decisions on what count as the proper conditions. There will always be some room for doubt.

We should even be careful in reporting what we see with our own eyes. The Sun sets in the west—but it doesn't really. According to modern science, it sits there while the western horizon rises. But this is exactly our project, connecting what we see to what we can conclude regarding the rotation of the Earth.

Surely there are some basic observations we can count on. Sparrows can fly; pigs cannot. Let's not descend into the debilitating and nonproductive full-on skepticism that is common to college sophomores and their enabling philosophy professors. Often under the influence of the sixteenth-century French philosopher René Descartes, the doubt goes all the way down to the most basic observations, even about sparrows and pigs. Descartes wrote at the time of scientific and religious confusion, the same time as Galileo, just after Copernicus. There was reason to think that scientists and scholars might have been wrong for millennia, and wrong on important issues such as the center of the world and the stability of the Earth. Descartes vowed never to be fooled again, and the only way to insure this would be

> I ought no less carefully to withhold my assent from matters which are not entirely certain and indubitable than from those which appear to me manifestly to be false, if I am able to find in each one some reason to doubt, this will suffice to justify my rejecting the whole.

Descartes was uncompromising in applying this method of doubt, clearing the slate of everything he previously believed—everything. He then worked hard to find the indubitable pieces of knowledge to put back in and build a description of nature beyond any possible doubt. He failed spectacularly. Reading Descartes today, it's easy to follow the first part of his project and acknowledge that nothing is certain, and just as easy to spot the mistakes in the second part, the attempt to rebuild from scratch. That's why many modern readers end up as skeptics.

Descartes' rescue mission for scientific knowledge was destined to fail from the beginning, when he set the threshold for accepting ideas impossibly high. He made remarkable contributions to physics and mathematics, but we are fortunate that it was Galileo's acknowledgement of the essential uncertainty that guided the method rather than Descartes' insistence on certainty.

Sparrows, though, I'm sure can fly. Pigs can't. I would stand by these truths before the Inquisition. But this is not about me, or Descartes, or even sparrows and pigs. It's about the Earth, and whether it rotates, a situation in which the reality is not immediately apparent.

Science is all about the difference between appearance and reality. The Earth is round, though it appears flat. A rock feels solid, but really it is mostly empty space of atomic nuclei and electrons. There are tiny disease-causing objects on your hands that are significantly removed by washing with soap and hot water, though none can be either seen or felt, and so on. In defying the apparent, science takes on a burden of proof. Bearing the burden is the responsibility of the scientific method, and there is a variety of techniques to connect the ways the world appears and the way it really is.

Sometimes it's simply a matter of changing the way we look at the world that brings appearance into agreement with reality. For example, it may be a matter of the proper conditions for viewing. Physics tells us that under the influence of gravity a heavy object falls to the ground at exactly the same rate as a light one, no matter how heavy or how light. The acceleration of gravity is a constant, independent of mass. That's the reality. But try it at home and you'll find that the results of a careful experiment defy this law of physics. Drop something heavy and something light, like a stone and a cotton ball, from the same height at the same time and the stone will hit the ground first. Aristotle did this and concluded that heavy things fall faster than light. He followed the evidence, albeit what we would consider pretty sloppy evidence, since he never measured how much faster. The problem, of course, is the air resistance. Only by doing the dropping in vacuum can we see the uninhibited effect of gravity. Do it on the Moon, or in an airless chamber, and the stone and cotton, like the Apollo 15 hammer and feather, hit the dirt simultaneously. Thus, under appropriate conditions, does the appearance match the reality.

This strategy won't work for the rotation of the Earth. It's a big system, too big to control the conditions in any way, let alone a way that would reveal the rotation, plain and simple. Controlling experimental conditions is generally out of the question in matters astronomical, the specimen being too big and usually too distant. Geology has similar challenges with unmanageably large objects and events far in the past. But there are other options for matching appearance to reality. One is to enhance the power of observation by way of instrumentation. Time-lapse photography shows that glaciers move, just as the glaciologists said they would. Microscopes reveal the germs on your unwashed hands—well, they reveal *something* microscopic on your hands, without actually showing it to be the cause of sickness. A telescope magnifies planets to show their Earth-like shape and orbiting moons. Some of the telescopic data were helpful to the Copernican cause. The moons of Jupiter showed that a moving planet could hold on to its orbiting satellites. The full spectrum of phases of Venus showed it had to orbit the Sun rather than the Earth. But no amount of magnification could show the rotation of the Earth. The Earth is plenty big and right here. If we could see it rotate, we wouldn't need a telescope or a microscope to do so.

There's a third option for bringing appearance into compliance with reality, and that is to reinterpret what we see. This is not changing the physical conditions of observation; it's changing the conceptual conditions, the theoretical context. Since different theories can provide different contexts and potentially different interpretations, observation starts to lose its objective, neutral authority. But really that loss started right away, as soon as there was any need to physically control or manipulate the evidence. What count as proper experimental conditions, and hence what can be used as a credible observation, may vary from one theory to another. What is allowable as a reliable tool for enhancing rather than distorting an image may depend on the theoretical context. Interpretation has been part of the process of scientific observation all along. Uncertainty is in it from the beginning.

The overt reinterpretation of the evidence, that is, not changing the image but only what you make of it and how you describe it, is generally less convincing to a skeptic than showing an enhanced or controlled image. Watch the Sun go down and listen to a pedantic physicist say that what you are *really* seeing is the horizon rise, and it still looks the same. We haven't changed our poetry, or the weather report, under the influence of the Copernican revolution. We still talk about and enjoy the sunset.

More often the scientific burden of proof is met not by enhancing or reinterpreting what is observed, but by using it as evidence for what is not observed. The connection is by inference, as in, if this is true then that is true. The legal term for this is circumstantial evidence. The contrast is direct evidence, such as a reliable witness to the crime, what in the context of science would be a direct observation. Circumstantial sounds suspect. It certainly is on television, with lawyers always objecting on the grounds of evidence being circumstantial, as if that automatically makes it unreliable. But circumstantial evidence is the empirical basis of scientific method, connecting hypothesis to evidence by way of credible inference, drawing justified conclusions about what cannot be observed on the basis of what is observed. The clearest examples are in medicine, diagnosing illness. There it's just called symptoms, and the routine is to determine the cause of a disease without actually seeing the pathogen. Fever, aches, chills, and so on are the circumstantial evidence of the flu. From the Surgeon General to a school nurse, medical science makes this kind of diagnosis all the time. Nobody objects to it as being circumstantial, not even on television, where there are at least as many programs featuring doctors as lawyers.

Sometimes the symptoms are suggestive but insufficient to narrow down the list of possible causes. That's when the doctor orders, in the clinical jargon, further tests, usually some way to isolate and magnify the microscopic cause of it all. This is the enhanced observation described previously. But in the majority of scientific cases, and almost all of the really interesting ones, no such enhanced observation is available. There is no isolating and magnifying the components of evolution, or the big bang, or the theorized dark matter the astrophysicists now say makes up the majority of cosmic mass. The burden of proof for these theories is carried by indirect evidence, circumstantial evidence.

In these terms of direct observation and indirect (circumstantial) evidence, what is the status of the rotation of the Earth? What was the empirical support for claiming the Earth rotates, or doesn't, in the ancient time of Aristotle, or more modern time of Galileo, or now? If it's not a matter of direct observation or enhanced observation, there has to be a logical link between what is observed and the conclusion that the Earth rotates.

A lot depends on the inference, the logic that connects observation to theory, symptoms to diagnosis. Consider the possibilities. It may be as straightforward as knowing that a particular phenomenon must happen if the theory under consideration, call it the hypothesis, is true. Cause and effect. Then look to nature for the phenomenon, or create the conditions in which it can happen, and see that it does. Diagnosing disease is a clear case like this. On the hypothesis that a patient has the flu, look for a fever, because the flu causes a fever. If it's the flu, there has to be a fever. But of course, the singular fact of a fever doesn't prove that the patient has the flu. There are lots of other diseases that present with a fever, so the one positive symptom does not narrow the diagnosis to any one of them.

The logic here is characteristic of science. A hypothesis is tested by its observable predictions, as in, If the hypothesis is true then such and such precise phenomenon will be observed. And just like the fever, finding the predicted phenomenon does not prove the hypothesis to be true. So, like a good doctor, a good scientist will keep looking, deriving multiple predictions of a variety of phenomena. As more independent predictions are observed to be true, the likelihood of the hypothesis increases. The goal is not a single, decisive piece of evidence, because that is impossible in this logical context. It's a preponderance of evidence that's being built. Here is our warning that the proof that the Earth rotates will not happen in a moment with a single decisive piece of evidence, that is, unless we can simply see the rotation directly. Unless there's an eye witness, the evidence will be circumstantial, and we will need more than one piece.

Before a hypothesis makes it into the textbooks as being no longer hypothetical and ready to pass along to the next generation of scientists, it should be beyond reasonable doubt. The confirmation of a hypothesis by indirect evidence will never be beyond all possible doubt. In some courtrooms there is a criterion of "reasonable scientific certainty" for the testimony of experts on evidence such as DNA analysis. No one really knows what "reasonable scientific certainty" means, including both the National Commission on Forensic Science and the American Bar Association. It's likely that most jurors hear it simply as "scientific certainty," unaware of the limiting logic of scientific testing and Galileo's admonition that those two things, science and certainty, just don't go together.

But we have only looked at one form of inference between evidence and conclusion. There are other possibilities. Maybe the hypothesis describes the *only possible* cause of some observable phenomenon. This is a much closer logical connection, and observing the phenomenon would prove, beyond any doubt whatsoever, that the hypothesis is true. Of course, it doesn't apply in a case of the flu, because it's just not true that the flu is the only possible cause of a fever. In fact, it doesn't apply in any real case in science. Nature is too complex and our understanding is too incomplete to make the causal connection so exclusive or so well-known. There are always multiple explanations for any observed events. To think otherwise reveals either a lack of imagination or an underestimate of the diversity in nature.

These are the reasons behind the disclaimer in the first chapter of most introductory science textbooks that a hypothesis can never be proven. They mean never proven with certainty, although the confident tone later in the text seems to forget the point. The can't-prove warning in the textbook account of scientific method is usually followed by a can-disprove declaration. A patient who does not have a fever cannot possibly have the flu. The doctor can rule that out and has to consider other hypotheses, diseases that do not cause fever. Even if there are hundreds of fever-causing diseases and hence hundreds of explanations of a fever (whether we know them all or not), there is no ambiguity in the negative test. It can't be any of these; it has to be something else.

In the language of the logicians, fever is a necessary condition of the flu. The effect must always happen when the cause is present. No flu without fever, so no fever proves no flu. This is what doctors do, at least in the preliminary steps of diagnosis; they rule things out. The implication in the textbooks is that this is what all scientists do, and pretty much all they do. Scientific method, they say, is a matter of falsification rather than confirmation.

There is something to this one-sided characterization of science, and the tidy logic is appealing. But we'll see in the case of the rotation of the Earth that the connection between evidence and hypothesis is not so simple. It never is. Our case is easier than most, in the sense that there are just two choices; either the Earth rotates or it doesn't. Disproving one would prove the other. If the one hypothesis cannot be proven, then it must be that the other cannot be disproven.

Aristotle followed the advice to try to disprove a hypothesis. He considered the possibility that the Earth rotates and derived the implications, the necessary conditions. Finding that the predicted phenomena did not occur, he concluded that the Earth does not rotate. No fever, no flu. We will follow the details of his argument in the Ancient Perspective, but at this point, note at least that his method matches exactly what the textbooks prescribe today. Note as well that Aristotle didn't presume the stability of the Earth was simply and directly observable. There was no eye witness; he relied on circumstantial evidence.

By the time of Galileo there was still no direct evidence of either the rotation of the Earth or its standing still. Nor was there an incontrovertible logical proof, or disproof, one way or the other. Reasonable people disagreed, including earnest and careful scientists, respectful of both evidence and reason. Lack of direct observation sustained the controversy, that and the cultural importance of locating the center of the world. There is no controversy any longer; now we are confident the Earth rotates. It's worth knowing the source of that confidence, and to understand its legitimacy. So we will follow the evidence, from ancient Greece to the present. When and how was it proven that the Earth rotates? Or, mindful of both the textbooks' admonition against the possibility of proof and Galileo's counsel on uncertainty, follow the preponderance of evidence to see when and why it became reasonable to believe the Earth rotates, maybe with reasonable scientific certainty. And there's the possibility that somewhere along the line, by some enhanced means of observation, the rotation became plainly visible. Maybe we'll find an eye witness.

Galileo defended himself and his *Dialogue Concerning the Two Chief World Systems* by claiming it did not favor either system over the other. The tone of the dialogue and the winner of the debate make this hard to believe. By the time of the trial, and for many years before, Galileo believed that the Sun was at rest in the center of the world while the Earth revolved and rotated. He wrote as much in 1597, in a letter to Johannes Kepler, the German astronomer who revised the Copernican model to allow elliptical planetary orbits. "Like you," Galileo wrote, "I accepted the Copernican position several years ago and discovered from thence the causes of many natural effects which are doubtless inexplicable by current theories." Kepler responded

> Be of good cheer, Galileo, and come out publicly. If I judge correctly, there are only a few of the distinguished mathematicians of Europe who would part company with us, so great is the power of truth.

Kepler, we now know, overestimated the persuasive power of truth. Galileo did come out publicly, in the thinly disguised dialogue with which the Inquisition would take issue. His conviction in the Copernican world system led to his conviction by the Catholic church. There's a rumor that Galileo had the last word at his trial, under his breath, after the reading of the verdict and the sentence and his own forced endorsement of the old world system. It may or may not be true, but the story goes that he muttered, *"Eppur si muove,"* and yet it moves.

The Ancient Perspective

CHAPTER 2
To Save the Phenomena

... trying by violence to bring the appearances into line with accounts and opinions of their own.
—**Aristotle**

Copernicus was not the first cosmologist to go on record with a model of the universe that put the Earth in motion. There were sober and literal-minded naturalist at least as early as the fifth century BC who claimed we stand on a rotating sphere. Saying the Earth is spinning was neither metaphorical nor poetic flourish. They were serious, and they were taken seriously. The rotating Earth had to fit into a larger description of the universe, and we need to be clear on what the rest of the universe is doing as the world turns. Equally important are the reasons given in support of this very counterintuitive notion that the Earth moves. In the context, the reasoning must not have been persuasive, as it would be another 2,000 years before the idea of a rotating Earth was added to the scientific canon.

If you're going to talk about the rotation of the Earth, you're going to have to say something about its shape. The ancient Greeks knew the Earth is round. This was more than a minority opinion or peripheral challenge to a majority doctrine of flatness. It was the generally accepted belief among scholars. From at least the time of Pythagoras in the sixth century BC, an education and interest in the natural world included the idea, again somewhat counterintuitive, that the Earth is round.

You've heard of Pythagoras as the mathematician who pointed out the relation among the sides of a right triangle and thus began the discipline of trigonometry. There was a lot more to his contribution to our understanding of the world, at least we think there was. He left nothing in writing, but his personality and ideas fostered a cohort of devoted followers that lasted more than a century. In this group of Pythagoreans, and in all scholarly activity since, there has been no suggestion that the Earth is anything but round. Some suggested the shape was a cylinder, but prevailing models were spherical.

It wasn't just a guess or a whim. There were explicit reasons to think the Earth is a sphere. Some of the challenges to determining the shape of the Earth are similar to those in figuring out if it moves. There was no external perspective on the object, no stepping off for a look back. That extraterrestrial view would not be possible until the twentieth century. And the Earth is huge, so the curvature is gentle, locally unobservable. As the Earth's motion is relatively slow, its size is relatively large. In both cases, the evidence would require deliberately gathered data that required interpretation. It would have to be circumstantial. The conclusion about the shape was unanimous and it hasn't changed. The ancient verdict on motion was a little more contentious.

It's worth a quick look at the ancient evidence that the Earth is round. Importantly, there was a variety of phenomena in support of the conclusion. That's how science works.

Pythagoreans and their contemporaries recognized that a lunar eclipse is caused by the shadow of the Earth cast upon the Moon. Eclipses differ in degree, some total but most partial. In all cases, the edge of the Earth's shadow is circular, never straight. This indicates that the Earth is round, and it favors a sphere over a cylinder.

Greek scholars often moved around their known world. Pythagoras, for example, was born on the island of Samos but moved to Croton in southern Italy. Mathematicians often traveled to Egypt where the discipline flourished. Some noted that the position of the fixed stars subtly shifted as an observer moved north or south, but not as one went east or west. There were stars visible in Alexandria that were hidden below the southern horizon in Athens. This would happen if the surface of the Earth is curved.

The data from eclipses and the stars are celestial, looking up to determine the shape under foot. There was also a terrestrial experience that revealed the curvature of the Earth. Ships sailing out to sea would be seen to slowly disappear with the hull first dropping below the horizon, followed gradually by the mast. It's as if the ship is sinking, but, of course, it isn't. It's following the curved surface of the sea. And, since this phenomenon happens equally in whatever direction the ship goes, a cylinder just won't do; it has to be a sphere.

There is an important difference to consider between the evidence for the shape and the evidence for the motion of the Earth. Shape is not relative. The Earth is not round *with respect to the Sun*, but flat *with respect to Spica*. There is no distinction between solar shape and sidereal shape. No reference is needed to determine— whether that means to define or to measure—the shape of the Earth. The modern term for this is intrinsic curvature, and intrinsic cur- vature can be measured using the features on the thing itself. The view of departing ships is one example. Another, not available to the Greeks, but simply because they lacked the logistical wherewithal for such long-distance travel, would be to continue in a straight line, in any direction, and you will end up where you started. This can't happen on a flat Earth. We could do this in an airplane. Or more practically, simply fly in a straight line—and that just means the shortest path—from one point to another. Turn and go straight to a third place, and then turn again and go back to the start. Measure the interior angles of the triangular route you just flew and you'll find they do not add up to 180°. That, by the way, doesn't happen on a cylinder, where the sum is always 180°, no matter how big the triangle or how it is oriented. So, modern aviation, like ancient navigation, offers some evidence that the Earth is a sphere.

The shape of the Earth was not a disputed issue in antiquity, neither in terms of it being relative nor having equivocal evidence. The motion of the Earth was a different matter. Following the ancient evidence for rotation should be done with an awareness of the sketchy availability of the primary sources. Greek scholars wrote their ideas on fragile materials such as papyrus, and most of it has been destroyed or lost. What we know of the Pythagoreans is through references by later Greek writers. Pythagorean reasoning about the shape of the Earth was retold by Aristotle. The cosmological models were described by Aristotle and others. Aristotle credited the Pythagoreans with these cosmological ideas, but we don't know if they were the first. Consequently, what's reported here may not be the the first suggestions that the Earth moves, or even the first on record. It's the first for which we have a record.

Why would anyone think the Earth rotates? There is nothing in anyone's immediate experience, the day-to-day encounter with nature that suggests it. The idea could only come from looking to objects in the sky and noting the apparent motion of the Sun, the Moon, the stars, and the planets. These celestial objects move across the sky, but very slowly. It requires some dedication to the task of observation and precision in the measurements to accumulate the data to support a model that puts the Earth in motion. Astronomy had to be the source of the evidence for a cosmology that moves the Earth.

There had long been a practical role for astronomy. The changing position of the Sun provides a reliable clock during the day. Sundials, in some cases just a stick planted vertically in the ground—a device called a gnomon—added some precision and repeatability to the data. They also facilitated a record of seasonal changes, as the length of the shadow correlates to the time of year. The one device served as both clock and calendar. Tracking more details in the night sky allowed for some fine-tuning of the calendar. The array of celestial objects shows up each night, and early astronomers assumed that the stars and planets continue to exist during day, even when they are invisible in the glare of the Sun. The pattern, identifiable by recognizable stars and constellations, is unchanged (with a few notable exceptions) but arrives just a little later each night. The delay is roughly 4 minutes each day, one degree of arc. This brings what looks like a dome of stars through one full cycle, 360°, in the course of a year.

The pattern of stars appears to move as a unit, as there is no relative motion among them, hence the reference in antiquity to the fixed stars. Think of it as a single sphere of embedded stars with a daily revolution from east to west. The Sun also goes around once a day, but just a little bit slower, so that it appears to drift eastward across the pattern of stars. Each day, the Sun falls about 1° behind, about 4 minutes in timing, so that after a year it's roughly 360° behind, one full circle. Each year, the Sun follows the same path through the fixed stars, the ecliptic. By choosing twelve constellations on the ecliptic, twelve periods of the year are identified. Thus, the location of the Sun corresponds to the month. Constructing a calendar this way uses the fact that the Sun appears to move in two separate ways, once around each day (a diurnal motion that accounts for day and night) and one circuit of the ecliptic each year.

Other objects in the sky are a little more complicated. These are the planets, and in antiquity the term applied not only to what we now identify as planets but also to the Moon and the Sun, anything in the sky that changes position relative to the fixed stars. Five planets other than the Sun and Moon were known, Mercury, Venus, Mars, Jupiter, and Saturn. Like the Sun, they move from east to west with the diurnal revolution and they drift slowly eastward along the ecliptic. The time to complete a full circuit around the ecliptic is in some cases much longer than a year. Saturn, for example, takes nearly 30 years to return to the same spot in the pattern of stars. And these planets have a curious tendency to sometimes stop in their slow eastward trip through the stars, turn back for a few days of westward motion, stop again, and continue eastward. The westward backtracking later became known as retrograde motion.

Carefully keeping track of these repetitive movements among the objects in the sky facilitated predictions of upcoming seasonal events and the scheduling of celebrations and holidays that were linked to celestial situations. Astronomy served for the benefit of agriculture, religion, and astrology. It served to the degree that it was accurate, and for the most part that was all anyone asked of the science, reliable forecasting and a steady calendar. Explanation of events in the heavens was not important, as long as the tool of description was sharp. And in this role, there was little need to link astronomy with cosmology, the understanding of the large-scale structure of the natural world and our place within it. There was generally no need to challenge the obvious sense that the Earth was standing still.

The exception was in the group of Pythagoreans, sponsors of a fairly comprehensive package of beliefs about the natural world and human existence. They made claims about the shape and structure of the universe, about the human soul that survived the death of the body, and even about the appropriate human diet. Pythagoreans were vegetarians, but they refused to eat beans; it's not clear why. Mathematics was the core of their description of nature, to the point that numbers were the essential building blocks of the physical world. Everything could be represented by numbers, and mathematical relations structured the world. Some numbers were better than others. A perfect square such as the number four was to be respected and it represented the noble concept of justice. Marriage was the number five, since it was the sum of a man, the number three, and a woman, two, and so on.

The numerical scheme was applied to objects in the sky. It was assumed that the stars are fixed onto a single sphere, the speckled dome we see at night. The planets are also on spheres, each to its own, which moved independently. This motion creates gentle sounds that blend together in harmony. When asked why we don't hear this harmony of the spheres, the answer was simple. It is unchanging and always present, in our ears since birth. We quickly and naturally come to ignore it.

Notable among Pythagoreans was Philolaus of Croton (470–385 BC). He is the first individual we can credit with a cosmological model that has the Earth in motion. His work is known only through references by later Greek and medieval writers, but the details are consistent and the sources are reputable. Copernicus cited Philolaus.

Two fundamental principles constrained Philolaus' description of the universe. The first was a Pythagorean reverence for fire. Looking ahead in the history, this metaphysical belief will show up again with Johannes Kepler who would insist on putting the Sun at one focus of the elliptical orbits of planets. In Latin, the word *focus* means hearth; *panis focaccias*, what we now call focaccia, is a flat bread baked directly on the floor of the oven, that is, on the hearth. For Philolaus, the importance of fire would put it at the center of the universe. The second key principle was the respect for numbers, meaningful numbers like perfect squares in particular. As a shamefully anachronistic aside, note how well this would work with the solar system as we understand it now. We put the fiery Sun in the middle and surround it with nine planets (before Pluto was downgraded). That's a perfect square. Pythagoreans would be pleased, and probably insist that Pluto be reinstated.

But Philolaus did not put the Sun at the center of the universe. He had an actual fire in that spot. With none of his own writing as reference, it is just speculation on the reasoning, but perhaps it was an association with the hearth as the center of a household, or the warmth of a womb as the source of life, or the mythological fire in Zeus' palace. Whatever the motivation, Philolaus made it clear that the central cosmic fire is never visible. He constructed a world system that demonstrated how this is possible, while allowing the phenomena that we do observe.

The invisible fire is the center of things. Around this, on concentric orbits, are the Earth, the Moon, the Sun, the five planets, and furthest out, the fixed stars. That's the snapshot; now put things in motion in a way to account for the basics like sunrise and set, the daily motion of the heavenly bodies, and the movement of planets through the ecliptic. The stars don't move; they are genuinely fixed stars. The Sun orbits the central fire once around per year. This accounts for its yearly trip along the ecliptic, through the constellations of the Zodiac. Between the central fire and the orbiting Sun, the Earth orbits the fire once a day. It also rotates once a day around an axis perpendicular to the plane of orbit. This means that one hemisphere of the Earth is always looking out away from the fire, while the other hemisphere is always looking in. We live on the hemisphere that looks out, and that is why we never see the central fire. As the Earth orbits the fire in one day, half of the orbit is spent on the same side as the Sun, and the other half is opposite the Sun—day and night. The other planets, Mercury through Saturn, orbit the central fire on separate spheres at different rates, their ride through the ecliptic.

This actually works. The Earth has two motions, both diurnal. We are sometimes turned toward the Sun and other times turned away. We are always facing away from the fire. It's like standing on a spinning carrousel, always facing out, with a bright light on the ground some distance away. The light sweeps across the panorama, showing up on one periphery (sunrise), moving evenly to a position straight out (noon), and then disappearing on the opposite side (sunset), to leave us in darkness but then to reappear and repeat. That happens once a day. The bright light itself is very slowly moving in a circle around the carrousel, in a way that changes its position against the distant background, returning after one full revolution. That takes 1 year. Thus do the Sun and planets drift along the ecliptic, the path through the distant background of the fixed stars. This accounts for all celestial phenomena except the retrograde motion. As a first approximation of a cosmological model, Philolaus' system had virtues.

There is one more piece in the model, an additional moving part. Count the components so far. The sphere of fixed stars is just one; it's the sphere that counts, not the individual occupants. Add in the five planets, and then the Sun, Moon, and Earth, and there are nine spheres around the central fire. That sounds great, nine being a perfect square, but to a Pythagorean, ten is even better. It's not just one better; ten is an ideal number, more important in the metaphysical scheme of things than a perfect square. There should be ten spheres in orbit around the central fire. This may be the reason Philolaus added the extra piece. Between the Earth and the fire, and so on the inner-most orbiting sphere, is the so-called counter-earth. Little is known about counter-earth, since it, like the fire, is never visible. It is always directly between the Earth and the fire, orbiting with exactly the same period as the Earth. Since we are always turned away from the fire—that's caused by the daily rotation of the Earth—we never see either the fire or anything in between.

Counter-earth seems to have been too much for Aristotle, adding details to the cosmological system to accommodate occult ideas about numbers, and then conveniently hiding them behind our back. This is where he accused the Pythagoreans of "trying by violence to bring the appearances into line with accounts and opinions of their own."

Any scientific theory will have to be a balance of common sense and consistency with observation. One group's common sense may be another's occult. Some of the basic principles of Philolaus' world system, the importance of fire and the governing role of numbers, made sense to the Pythagoreans but were not common among other schools of thought. The fire at the center of things and the counter-earth were idle components in the model, having nothing to do with the evidence, the appearances, as Aristotle put it. Why not just remove them from the system, not just to make the theory simpler but to eliminate what cannot be detected? But that leaves an empty space in the center of the universe, and a void point hardly seems up to the task of centering the rotations of the remaining nine spheres. The next step, then, seems almost irresistible. Move the Earth into that open spot at the center of the universe. This is exactly what would happen in a successor cosmological proposal, that of Heraclides Pontikos (390–310 BC).

Before we leave Philolaus, note that there was more that troubled Aristotle in his commentary on the Pythagorean system. It's not just the embellishment of experience with the addition of the central fire and the counter-earth; it's the more egregious defiance of experience in proposing that the Earth moves. This is the real violence. There is absolutely no sensation on the Earth itself of its being in motion. The terrestrial evidence is not neutral regarding the rotation of the Earth, as it may be about souls and mystic numbers. It tells unequivocally against. And yet scholars such as Aristotle must have taken the idea seriously enough to explicitly bring it up and dignify it with a response. The rotation of the Earth was neither ignored nor dismissed.

Heraclides Pontikos was a contemporary of Aristotle's and a fellow student at Plato's academy. He proposed a world system that is simpler than the Pythagoreans' and with less metaphysical baggage. Consequently, it comes with less explicit support for the details. Philolaus at least provided some theoretical and metaphysical background for us to reconstruct his reasons for thinking the center of the universe was occupied by fire, but without the context of a larger worldview, the placement of the pieces in Heraclides' model may seem somewhat arbitrary. Small fragments of Heraclides' writing survive, so later scientists could glimpse some bits and pieces of this thinking, but the dots are scattered and difficult to connect. Like most of the ancient Greek scholars about whom we know, his interests were extensive and eclectic, including politics, art, ethics, and cosmology. His ideas on the position and motion of the Earth are known primarily through reference by later astronomers, but there is no mention of Heraclides by Copernicus.

As the name indicates, he came from Heraclea Pontikos, a Greek town in Asia Minor, on the Black Sea. Apparently, Heraclides was something of a dandy. Rich and well-fed, his clothing was ostentatious and his demeanor aristocratic. This behavior, and the name of his home town, burdened him with the nickname Pompikos. In Greek it means stately and magnificent, and pompous.

Here are the basics of Heraclides' model of the cosmos and the situation of the Earth, as reported by subsequent writers. The Earth is round and at the center of the universe. The celestial bodies, that is, the Sun, the Moon, the five lesser planets, and finally the fixed stars, are all situated on concentric spheres around the Earth. These aspects of Heraclides' world system, the snapshot before anything is put in motion, are identical to what would be adopted by Aristotle and tinkered with for almost two millennia. Call it the standard model. But when things move, there is a difference. In the Aristotelian standard model, the Earth stands still; in Heraclides' model, the Earth at the center of the universe rotates on its central axis, once around each day. One way to look at it is to simply move the Earth into the central place of the fire in the Philolaus model (eliminating the useless counter-earth), while retaining the diurnal rotation of the Earth. There is no evidence that this is how Heraclides got to the idea, but from the distance of history, the progression is that simple. The fixed stars, as with Philolaus, do not move. The Sun, Moon, and other planets drift eastward through the ecliptic by slowly revolving around the central Earth. Day and night, sunrise and sunset, and the westward track of the stars across the night sky are all the apparent result of the Earth rotating to the east.

The clearest description of this Heraclides model is by Simplicius, a sixth-century AD commentator on many things Aristotelian and likely the model for the character Simplicio in Galileo's *Dialogue*. "Heraclides supposed that the earth is in the center and rotates while the heaven is at rest, and he thought by this supposition to save the phenomena."

In the fragmented record of Heraclides' thoughts, there is no indication of the evidence or reasoning to support the claim that the Earth rotates. We can speculate, but that's all it is, imposing our own way of thinking from a context 2,000 years after the fact. The Heraclides system is simpler than that of Philolaus; it eliminates the two superfluous (from our perspective) objects, the fire and the counter-earth, and it reduces the the number of motions of the Earth from two to one. With Heraclides, the Earth has no orbital revolution, only the daily rotation. In fact, every object in the system has just one motion, with the exception of the sphere of fixed stars that does not move at all. This makes the Heraclides model arguably simpler than what was more commonly accepted at the time, what I've called the standard model. That has the same structure but keeps the Earth still while everything else orbits once a day. Not only does this require two motions for the Sun (and each of the other planets)—the diurnal orbit plus the slower trip through the ecliptic—it also puts the celestial sphere of fixed stars in daily orbit. If the celestial sphere is large, it will have to move really fast to get all the way around in 24 hours. The speeds become greater as the distance to the stars is increased, and it is almost unimaginable if the universe is infinite, as Heraclides claimed. This was another break from the cosmological canon of his time. An infinite universe is another possible reason to put the Earth in motion and keep the fixed stars stationary, although the details are unclear and so this account of the motive is speculation.

The phrase "to save the phenomena" in the Simplicius description of Heraclides' cosmology was a common choice of words for describing the match between theory and evidence, common, at least from about the first century BC. It still is, among philosophers of science. To praise a theory for its success in saving the phenomena is to say that nature behaves as if the theory is true, but that doesn't mean it is true. The phrase "to save the phenomena" has become a way to call attention to the difference between *as if* and *is*. In this light, theories and models, including cosmological models, are valued only for their empirical success, with no claims made one way or the other regarding their being true. The idea is that science is a pragmatic business, and any model that saves the phenomena, that matches what has been observed and accurately predicts what will be observed, is acceptable. Thus, Heraclides proposed that the Earth rotates as a way to save the phenomena, but Simplicius is suggesting he didn't mean the Earth *really* rotates. It's just a way to think about the situation, a calculating device that works.

The merely pragmatic interpretation of a rotating-Earth cosmology can't be true of Philolaus. The Pythagoreans, serious in their metaphysical and religious commitments, put in more pieces than were strictly necessary to save the phenomena. This indicates a belief that the cosmological system is true and the Earth really moves. Heraclides' attitude is less clear. Rotating the Earth may have been just another way to look at the cosmos, an equally effective alternative to the standard model that has the Earth at rest. He simply gave one of the Sun's two motions to the Earth (or from our modern perspective, one of the Earth's two motions to the Sun), and either way saves the phenomena. This is a preview of the relativity of description of motion. Choose the Earth as the stationary reference point, or choose the celestial sphere of fixed stars, and you will be working with the standard model or the Heraclides model, respectively. Either way, the celestial objects are observed to behave as the model predicts. Either way, the phenomena are saved.

As long as astronomy is valued only for its catalog of heavenly objects and their changing positions, and for the resulting predictions of important seasonal and celebratory events, it doesn't matter whether the Earth rotates or not, as long as the phenomena are saved. Whether Heraclides adopted this utilitarian attitude is just a guess. Later astronomers seemed to have thought he did. Regardless, much of the discussion at the time regarding the rotation of the Earth reflects a sense that the proposed models were taking literally and at full realistic value. There was a serious claim that the Earth moves, and a genuine need to refute it.

The systematic refutation will be conducted by Aristotle in the next chapter, so it may be premature to decide, with Heraclides' theory on the table, if it would have been reasonable to believe that the Earth rotates. Nonetheless, it's worth raising the question. In light of the best available evidence, and with the help of the best available theories to interpret the evidence, how should a jury decide? Scientific decisions must be made in the context of the time, without the benefit of a perspective from the future. In the context of fourth-century BC Greek science, there were three theories in play to describe the large-scale structure of nature. In two of them the Earth rotates. All three matched the celestial data with equal, though imperfect, success. But the theories of a rotating Earth faced an obvious challenge closer to home. Consider the terrestrial data. No sensations here on *terra firma* indicate motion of any kind. A rotating Earth may save the celestial phenomena, but it does not save the terrestrial phenomena, so by any standard of scientific success, it must be rejected. This will be Aristotle's case.

CHAPTER 3
Aristotle's Standard Model

To be a good investigator a man must be alive to the objections inherent in the genus of his subject, an awareness which is the result of having studied all its differentiae.
—Aristotle

Credit Aristotle in the fourth century BC for taking seriously the suggestion that the Earth rotates. He had to ignore the obvious—the comforting stability of *terra firma*—and consider the absurd. His own conclusion, that the Earth is absolutely motionless, would seem to bear no burden of proof, but he took on the challenge nonetheless. That's good science. Unlike in a court of law, both sides in a scientific debate must make their case.

We should follow Aristotle's advice, alive to the objections inherent in what we now take for granted. That is, we need to understand his arguments that the Earth does not rotate. This is more than an interest in being fair and even-handed; it's the recognition that the Aristotelian model of the universe and its companion scientific method, the interrelation between theory and evidence, were the dominating standard for cosmology and physics from late antiquity, through the middle ages, and up to Copernicus. It's the elephant in the room, if you are interested in the evidence and reasons to believe the Earth moves.

Aristotelian science was a magnificently integrated and coherent system of detailed descriptions and explanations of a variety of aspects of nature. The physics of what happens on the Earth and the astronomy of the sky were consistently put together into a cosmology of the universe. The pieces cannot be understood or evaluated in isolation, nor could the worldview be challenged piecemeal; the ideas do not stand (or fall) alone. This would explain its longevity. Change had to be wholesale, replacing the entire comprehensive account of Earth and sky with another, equally consistent with the evidence. Replacing the Aristotelian system wasn't the result of accumulating more and better empirical data so much as it was a group effort of theoreticians, from Copernicus to Newton, who put together the pieces of a new model. The process of changing physics and the model of the universe was fraught. It's a good example of what we now call a paradigm shift.

Aristotle's authority lasted nearly 2,000 years with some tinkering and a few additional details but no fundamental changes. It was scientific canon, the standard model of the universe, with a stable core. The Earth, at the center of the cosmos, held still.

We know a lot about Aristotle. His enduring work is still actively studied and taught by philosophers, and his setting in classical Greece is described and analyzed in much detail by historians, archaeologist, and philologists. Remarkable things were happening. With the benefit of mathematical insights and careful astronomical records from Egypt and Babylon, Greek scholars conducted ambitious inquiry of nature, the universe, and the human condition. Prosperity and slavery allowed a cultured class of men to devote their ample free time to impractical pursuit of wisdom. The egalitarian spirit that shows up in the political ideals of democracy (exclusive of slaves and women) worked to the advantage of science as well. A free discussion of ideas and willingness to consider dissenting opinion can only help in the effort to understand nature. This is not to be confused with relativism, the notion that anyone's opinion is their own personal truth. No, the give-and-take in the scientific dialogue is meant to find the truth about nature by revealing mistakes while rewarding logical consistency and empirical success.

The style of doing science in consideration of alternative hypotheses resulted in an enduring record of who said what. We know about Philolaus in part through the references in Aristotle. And we know about Aristotle's natural theories both by what he wrote himself and what others wrote about him. A dedicated association of scholars developed around Aristotle and continued long after his death. As there were Pythagoreans, there were Aristotelians, less like a cult but much longer in existence. Participants in the Aristotelian school of thought were sometimes known as Peripatetics, Greek for those who walk, pedestrians. The reference is probably to an exercise field and walking trail in the public space next to the school Aristotle established in Athens.

It's estimated that Aristotle wrote nearly two hundred individual treatises. Thirty-one of these have survived and are available to be read today. The topics cover a general-subject list in a library, from physics and biology, to poetry and art, ethics, and politics. What we have are not the original physical objects touched and inscribed by Aristotle himself. Rather, they are medieval manuscripts, translated and transcribed by monks in what we trust is a tradition of fidelity and accuracy.

Aristotle was born in Macedonia in 384 BC, the late-classical period of ancient Greece. He missed the opportunity to study with Socrates, but at the age of 17 he was sent to Athens and became a student of Plato. He stayed with the Academy for 20 years, until Plato's death in 347. Aristotle left Athens and spent a few years in Asia Minor, and then the island of Lesbos. This itinerate life didn't last long, and in 343, Philip of Macedonia summoned the scholar back to his home to tutor the king's son, Alexander. After 2 years together, teacher and student went separate ways. Fifteen-year-old Alexander went on to greatness and to conquer as much of the world as he could; Aristotle returned to Athens.

Back in Athens, Aristotle established his own school, following the model of Plato's Academy. This is the one near a footpath. Since it was in a public area dedicated to Apollo Lyceios, the school became known as the Lyceum. You can still visit the area and walk through remains of the structure. It's next to the Athens Intercontinental Hotel, right downtown. Members of the Lyceum were notable in collecting and protecting manuscripts of scholarly treatises. Their archives became one of the first libraries.

Classical Greece was a collection of autonomous city-states that didn't always get along. Aristotle was not a native Athenian, and he left the city and his Lyceum in 323, apparently because of escalating anti-Macedonian activities. He moved to Chalcis on the island of Euboea, where he died 1 year later.

In the large catalog of works by Aristotle, two are directly relevant to the rotation of the Earth. The *Physics* has a lot to say about how things move, and *On the Heavens* clarifies the location and status of the Earth in the universe. The two treatises—you can't really call them books, since this is centuries before the invention of the printing press and the publication of books—and the separation of topics highlight the distinction between the two perspectives for evidence on the rotation of the Earth. Aristotelian physics is the science of what happens on the Earth. Astronomy is about the heavens. One will report on terrestrial phenomena, the other celestial. They will be different, but they will have to be consistent.

Physics, for an Aristotelian, is fundamentally the study of motion. This fits nicely with the notion of being peripatetic, but that's probably just a coincidence. More to the point is an often repeated Aristotelian aphorism: "To be ignorant of motion is to be ignorant of nature." The reference to motion is about change in general, as the growth from acorn into oak tree is a kind of motion, and a falling stone is another. What we think of as motion, moving through space, Aristotle describes as locomotion, change of location. There is plenty of information about locomotion in the *Physics*. Not so different from physics today, the Aristotelian analysis is about the composition of things and the movement of both the parts and the whole. There's a connection between what something is made of and how it moves, and this is relevant to the rotation of the Earth. Once we figure out what the Earth is made of, we'll know whether it is capable of rotation, or motion of any kind. Aristotelian physics was about laws of nature, discoveries of not simply what does happen but what must happen or couldn't happen in nature. The same is true of modern physics. And sometimes a clear application of the laws makes evidence all but unnecessary.

Science is all about distinctions and categories and types—species and genus, planets and stars, solids and liquids, and so on. Aristotelian physics starts with a distinction between two types of (loco) motion, natural and violent. These categories were discovered through careful observation. Natural motion is what happens when an object is left alone, uninhibited and unmoved by any external influences. The clear case is a dropped stone. Once you let it go, the stone falls on its own, and we can see that it goes straight down. That's natural, in the sense that it's in the nature of the stone to fall. Violent motion, or as Aristotle sometimes called it, enforced motion, is the result of something pushing or pulling or holding back on the thing. It's not just a human action that can cause violent motion; any external influence will do. Lifting the stone up is a kind of violent motion, as is tossing it. A stone rolled around by strong waves is violence as well. Despite the colorful language of violence, Aristotle's analysis of types of motion is not so different from what you'll find in a physics textbook today. There the key term is "free-particle." That's an object free of any external forces and consequently left to follow its natural trajectory through space and time. With no forces acting, inertia determines the motion, a straight line with no change in speed.

Beyond this basic distinction between forced and free particles, the details in Aristotle's account of how things move are very unlike modern mechanics. Aristotelian natural motion is dependent on the composition of the thing, and on its location in the universe. It's not just about where it is, but where it belongs, and that depends on what it's made of.

All material things we encounter on the Earth are made of some blend of the four elements, earth, water, air, and fire. The stone, for example, is mostly earth, with probably a little bit of water and air. A piece of wood will have some earth in it, but also a good deal of water and considerable amount of air. Sea foam, of course, contains water and a generous amount of air. Of the four elements, each has a proper place in the universe, a place where it belongs, and when left alone, it will seek out that place. This is natural motion. The dropped stone falls because the proper place for the element earth is to be below air. The same stone sinks because earth belongs beneath water. The natural order of the elements has earth at the bottom, then water, air, and fire on top. Natural motion is the process of restoring this natural order to the universe, to undo the disorder that has been caused by violent action such as your lifting the stone off the ground. Without the ongoing violent actions on the Earth, including human and nonhuman activities, the natural order of elements would exist in pure layers of earth, water, air, and fire. We'll see in a moment why these would be in concentric rings.

The natural sorting of elements is vertical because the natural motion of each element is vertical. All natural motion is in a straight line, either up or down. Earth, the element, moves straight down, as does water. Air and fire move straight up. And since each element has exactly one proper place in the universe, every material object will have exactly one natural motion. Free from unnatural interference it will move vertically up or down, depending on its composition, and with a speed determined by the amount of each element it contains. This is the general law covering all things that move on the Earth.

Natural motion is different in the heavens. Astronomical observations reveal that celestial objects and phenomena are eternal and repetitive. Nothing changes in the night sky, other than the graceful cyclical rotation of the fixed stars and the steady migration of planets along the ecliptic. The same stars return each night; none disappear or change, nothing new appears. There is none of the messy disruptions, the episodic birth and death, destruction and rearranging that take place on Earth. The celestial realm is altogether different from the terrestrial, and so it must abide by different laws.

By observing the orderly activities of celestial objects, Aristotle concluded that there simply is no violent motion in the heavens. It's all natural. But celestial natural motion is obviously not straight or vertical; it's not directed toward or away from the center. It's circular. And this makes sense, since a circle is the most perfect, uniform, symmetric shape, and circular motion is repetitive and eternal, as is seen in the stars and planets. And, since the details of natural motion are determined by the elemental composition of the body, celestial objects must be made of different stuff. It can't be earth or water, since any amount of either of these would cause the stars to fall. Nor can there be air or fire in the heavens, for then the stars would flee from the center. There must be a fifth element, the nature of which is to circle the center. Aristotle called it aether. In later translations it is sometimes given a Latin name, quintessence, the fifth element.

Made of different elements and consequently moving in different ways, celestial and terrestrial things must abide by fundamentally different laws. But the two sets of laws must be compatible, since somewhere there must be an interface, where Earthly matter rubs up against the heavens. Aristotle did not allow for any empty space that would isolate the two realms. Nature does not accommodate a vacuum. The Moon, or just beneath it, marked the boundary. Anything below the Moon was of this Earth. That's the space for the atmosphere and the clouds. It is also, by Aristotle's reckoning, where comets came and went. He discussed comets in a treatise on meteorology. Putting comets below the Moon, in the realm of violent motion and corruptible phenomena, would factor into later arguments about the rotation of the Earth.

Putting these pieces of physics together with the latest astronomical data, Aristotle assembled a cosmology that would be the core understanding of the universe to last for two millennia, what I have been calling the standard model. It helps to understand and evaluate it in the context of a challenge attributed to Plato, Aristotle's teacher: "What are the uniform and ordered movements by the assumption of which the apparent movements of the planets can be accounted for?"

The "uniform and ordered" is generally interpreted to mean circular. Build a model of the cosmos that includes the fixed stars, the planets, and the Earth (since that's the vantage from which the apparent movements of the planets occur) that has celestial objects moving on circular trajectories and that results in a match with astronomical data. It will have to account for the daily rising and setting of the pattern in the sky, and the more leisurely planetary drift along the ecliptic.

The first response to Plato's call to cosmology that we know with some detail is by Eudoxus of Cnidus (ca. 400–350 BC). He devised a model of twenty-seven concentric spheres with the Earth stationary at the center. Each planet was on a system of three or four spheres with off-set axes to rotate in ways that produced both the diurnal and ecliptic motion. It was complicated, and it's not clear whether Eudoxus intended it as a real replica of the universe or more humbly as a calculating device, a model that would save the phenomena. In either case, it's complication was likely its downfall, and it was soon replaced by the Aristotelian solution.

Aristotle pretty clearly regarded his own cosmology as the real world, not metaphor or mathematical tool but a description of the way things really are. Following the laws of physics and the mechanics of motion, he pointed out that the fundamental directions of up and down, and hence the physical properties of being heavy or light, would be meaningless in a universe of infinite extent. There must be a central point that defines the direction toward which heavy things fall and from which light things rise. Thus, the universe is finite; infinity has no midpoint. The universal center point determines the natural motion of things. Terrestrial natural motion is radial, that is, along a straight line from the center out toward the edge. Celestial natural motion is tangential, around on a circle concentric with the universe itself.

There is nothing outside the finite universe, nothing beyond the celestial sphere of the fixed stars. It's not empty space, again, since nature disallows a vacuum. This may be hard to visualize and understand, a boundary beyond which there is nothing, but really there is no cosmology without some detail that challenges the imagination. Whether it's bounded or boundless, eternal or with a moment that time begins, there is always something that defies explanation on familiar terms.

Once inside the celestial sphere, the details of Aristotle's cosmological model are easier to picture. Each planet is fixed on its own sphere and moves around the center point of the universe as the sphere rotates. These are real spheres, made of real stuff, the quintessence, the fifth element. These are the same celestial spheres that would be the topic of Copernicus' *On the Revolutions of the Heavenly Spheres*. Aristotle took them seriously, and so did Copernicus.

All of these spheres, the celestial sphere of fixed stars and each of the planetary spheres, orbit the center point of the universe once every 24 hours. This accounts for the diurnal east-to-west motion of the grand pattern of celestial objects, the rising and setting of the Sun and stars and planets. Each planetary sphere has a second, independent rotation in the opposite direction, as we record in the slow movement along the ecliptic. A picturesque analogy of the Aristotelian planetary system was provided by Vitruvius, a first-century BC architect and scientist.

> If seven ants were to be placed on a potter's wheel, and as many channels were to be made around the center of the wheel, growing in size from the smallest to the outermost, and the ants were forced to make a circuit in these channels, then as the wheel was spun in the opposite direction …

… the motion of the ants would resemble that of the planets.

Notice that there has been no mention of the Earth. The Earth is not the defining object in the standard model, but it has to fit in somewhere. We know from independent evidence like the shadow it casts on the Moon that the Earth is round. Aristotle's system has an explanation of why it's round. Keep track of the difference between earth the element and the Earth, our spherical home. Earth (the element) moves naturally to the center of the universe. That's just another way of saying it falls straight down. As a result, as things are stirred up, earthen components accumulate around that one point, forming a sphere. The solid Earth, being made mostly of earth, formed as a symmetric ball, centered on the point at the center of the universe. So, the natural motion of earth explains two things, why the Earth is round and why it is located at the center of the universe. In Aristotle's own words

> It so happens that the earth and the Universe have the same centre, for the heavy bodies do move also toward the center of the earth, yet only incidentally, because it has its center at the center of the Universe.

It's not that the Earth *is* the center of the universe. It just happens to be at the center of the universe.

The natural motion of earth explains something else, why the Earth is motionless. Again, as it is composed of earth, and those components reside naturally at the center, it would be unnatural for the Earth to move in any manner away from the center of the universe. Aristotle, again,

> If then any particular portion is incapable of moving from the centre, it is clear that the earth itself as a whole is still more incapable, since it is natural for the whole to be in the place towards which the part has a natural motion.

So not only does the Earth seem to sit still, it *must* be stationary, by law of nature. The shape, the position, and the stability of the Earth are all the necessary result of the law of natural motion.

This argument for the stability of the Earth works effectively against the Pythagorean claim that we inhabit a planet that orbits a central fire. That has the Earth off-center and moving around a circle. The argument also counters the Heraclides system, but less directly. For Heraclides, the Earth (and all the earth) are properly at the center, but the the pieces move unnaturally around, tangent to the center, rather than the singularly straight down along the spherical radius. That, too, is mechanically impossible. Thus do the laws of physics preclude the rotation of the Earth.

Turning to the celestial realm, Aristotle might well have titled his work on cosmology, *On the Revolutions of the Heavenly Spheres*. He was less descriptive, simply, *On the Heavens*. There is the central, stationary, messed-up Earth, surrounded by real, invisible concentric spheres that hold the stars and planets and that revolve, each at their own pace. Ants on a spinning potter's wheel. There are no gaps between the spheres, since there is no allowance for vacuum. This detail forced Aristotle to included intermediate spheres to buffer the motion and allow the independent rates of rotation. In all, the Aristotelian universe required 47 moving spheres. It's theoretical coherence was impressive, but the number of moving parts was troubling. And even with all those spheres in motion, it failed to account for the retrograde motion of planets, or the observed fact that their brightness varies, as if they are sometimes closer and other times farther away.

Despite having explained not only *that* the Earth is motionless but why it *has to be* motionless, Aristotle explicitly considered alternative arguments, both for and against the rotation of the Earth. Some of the evidence for a stationary Earth he found insufficient. One theory of the origin of the cosmos, for example, described the events of a universal vortex that sorted the elements and left the Earth at the center. The swirling universe left the debris in the center—not a very flattering description of the Earth—and sent the more tenuous elements to the outer edge. Comparing the effect to what you see in a cupful of spinning water, the center is motionless. In fact, this is something you can do at home, with paradoxical results. Make a cup of tea with loose tea leaves in the cup. Stir the tea and watch what happens. As the liquid circles around, the solid leaves accumulate in the middle and, at the very center, stop moving. This is the vortex model of a universe with revolving heavens and an unmoving Earth at the center.

Aristotle was unconvinced by the argument, even though he agreed with the conclusion that the Earth at the center does not rotate. Once again it is the fundamentals of physics that set the constraints on his cosmology. He argued that it makes no sense to credit a primordial vortex with sorting heavy things from light. The properties of heavy and light are defined in terms of a point in space, the center of the universe. That means that the center point must have been determined prior to any whirlpool, and consequently the accumulation of the solids would have happened independent of a vortex. Furthermore, there is no answer to the question of what happens when the constraint of the vortex is absent, as Aristotle argues it must be now that the Earth is separated from the heavens. He also thought the whirlpool provided no explanation for the upward motion of fire.

This is typical of Aristotle, arguing from fundamental principles and logic rather than paying close attention to the details of an experiment. He missed the mistake in the vortex analogy, the tea leaves swirling in a cup. They do gather at the center as the liquid circles around, but it's only on the bottom of the cup. The explanation, first made clear by Albert Einstein, is that friction between the rotating liquid and the stationary bottom of the cup slows the rotation at that interface and thereby reduces the centrifugal force on objects at the bottom. This creates a current of tea flowing from center to rim at the top where the rotation and outward force are strongest, and down the sides and back into the center at the bottom. It only works in a vortex that is walled-in on the outer rim and bottom. So, unless the universe is in a tea cup (and has been since the beginning), the analogy fails.

The conceptual misunderstanding that Aristotle saw at the core of the vortex theory of the stability of the Earth is the assumption that an active constraint is required to hold the Earth still, as if the stability is not natural but enforced. This, he said, is misguided. Since "the order of the world is eternal," both the position and the stability of the Earth must be natural, not forced. It would be any movement of the Earth, either displacing it from the center or rotating it around the center, which would require an active force.

These Aristotelian arguments address the cause of the Earth being motionless by pointing out the theoretical imperative. It simply *can't* move; it's impossible. There is a separate question of the evidence. It's almost unnecessary, given both the obvious stability of the Earth and the consistent theoretical reasoning. Nonetheless, Aristotle provided the empirical proof. He had evidence from both the celestial and terrestrial perspectives that the Earth does not rotate.

Consider the astronomical data. Aristotle believed they tell against the rotation of the Earth.

> If the Earth moved, there would have to be passings and turnings of the fixed stars. Yet these are not observed to take place: the same stars always rise and set at the same places on the earth.

It's not so clear what Aristotle meant by the passings and turnings of the fixed stars. It's likely he is considering the hypothesis that the Earth is like other planets, with two motions, diurnal rotation on its axis and annual orbit around the center of the universe. And he seems to assume that the axis of rotation must be aligned with the axis of orbit. This would cause a small seasonal change in the positioning of the fixed stars. The pole star, for example, aligned with the axis of rotation, would describe a slow, annual circular trajectory as the Earth's orbit carried the rotational axis around in a circle. The general pattern of stars would change orientation in the same way. But this does not happen, so the hypothesis is rejected.

Even allowing for what seems like a gratuitous assumption that the two rotational axes are parallel, Aristotle's evidence does not challenge the Heraclides model of the cosmos with the Earth rotating at the center. He clearly thought it does, "whether [the Earth] move around the center or is situated at it." But for Heraclides, the Earth rotates steadily with respect to the stars, the rotational axis always aligned with the pole star. The planets circle on independently tipped orbits to carry them along the ecliptic. The absence of seasonal change in the fixed stars may count against the Earth in orbit, but not against rotation, and as the Heraclides model demonstrates, it's possible for the Earth to have just the one motion, rotation. The celestial perspective does not rule out the rotation of the Earth.

The terrestrial evidence, on the other hand, carefully observing phenomena down on the Earth itself, does seem to rule out the possibility of a rotating Earth. Aristotle was brief in presenting this case, perhaps because it retells the obvious. He simply reminds us that, "heavy objects, if thrown forcibly upward in a straight line, come back to their starting-place, even if the force hurls them to an unlimited distance." Put this experiment in its context and interpret the results in light of the best scientific theory at the time. That's how science works. Throwing a stone straight up begins with an act of violence, your hand forcing the earthen object away from its proper place. As soon as you let go, the stone will abide by its natural motion that is straight and strictly vertical. If the ground is moving horizontally as the Earth rotates west to east, the starting place of the throw will go with the Earth and move eastward, out from under the falling stone. If the Earth rotates, the stone will fall some distance to the west. But, as we observe, that's not what happens.

Tossing the stone is just one example of many similar experiences that show the Earth to be stationary. It has the virtue of great height, the "unlimited distance" that would give the Earth sufficient time to move far enough to make the effect large enough to detect. But presumably, by simply hopping up in the air you would land a little to the west. Aristotle was making precise the common sense of stability on the Earth. And he was doing it with the tidy scientific logic of falsification. If the hypothesis of rotation is true, the prediction is to fall some measurable distance to the west. The prediction is repeatedly observed to be false, so the hypothesis is rejected.

Aristotle summarized the scientific data: "from these considerations it is clear that the earth does not move." Some of the considerations are theoretical, in the sense that they rely on the larger scientific understanding of nature to rule that a rotating Earth is impossible. It would violate laws of nature. We really don't have to look at anything to see if the Earth does move, because we know that it can't possibly move. But we look anyway, and some of the considerations are empirical. Of course, even the evidence bears some influence from the current theories of the physical world, most importantly the physics of motion. It couldn't be otherwise. If you turn to science for answers, you are going to get results that draw on the best and broadest understanding of nature at the time. In fourth-century Greece that would be Aristotle.

With the Earth rotating at the center of the universe, Heraclides is sometimes interpreted as only suggesting an alternative perspective, equally capable of saving the astronomical phenomena. It's an early variety of the relativity of rotation. Take it either way. The stars orbit relative to the Earth, or the Earth rotates relative to the stars; it's just an optional matter of perspective and choice of reference frame. Can Aristotle's scientific system accommodate this kind of relativity?

From the most foundational beginning, Aristotelian physics operates in an absolute space and consequently a framework for absolute motion. There can be only one descriptive perspective that represents the real position and motion of things. The system is built and oriented around a single point in space, not an object. This point, the center of the universe, is not relative to any material thing. Consequently, position in the universe is determined by position in space, relative only to the point. Something is either at the center, or some particular distance from the center, absolutely. And this establishes a system of absolute directions in space that are not relative to any object and not subject to changing perspective. Along lines radiating from the center the directions are up and down, the natural motion of terrestrial elements. Perpendicular to this, the tangential curves that circle around the center, is the natural celestial motion. None of these fundamental directions or properties is relative to objects or perspective. It is all determined by the centering point of space itself.

With this understanding of the nature of space, it cannot be left to an optional choice of reference frame whether the Earth is in the center of the universe or whether it is in orbit around the center. Either it is or it isn't, really. The Pythagorean cosmology with the Earth revolving around a central fire does not offer just a different perspective; it describes a different world, one that, by Aristotelian laws of nature, can't be the real one.

The location of the Earth is not a matter of perspective or choice of one's reference. Nor is the distance from the center relative to anything other than the spatial point itself. But rotation of an object at the center is trickier, and the Heraclides-inspired relativity may be allowed in the Aristotle's space. Almost surely Aristotle had no such intention, but the basic spatial properties accommodate the relativity of rotation. It's because the whole thing is anchored by a single point, the center of the universe. The rest of the universe is then spherically symmetric such that rotation around the center is determined only by reference to other things, not by any abstract spatial properties. This means that, while a rotating Earth may be ruled out by the mechanics of motion and the distinction of natural and enforced motion, it is not impossible in the more fundamental structure of space. Another way to look at it, it's only when we consider the cause of the motion that a rotating Earth is ruled out; the motion itself is relative. With a different theory of the causes of motion, a rotating Earth might be possible. Aristotle was surely no relativist. Heraclides might not have been either. But it's helpful to know just how deeply into the theoretical system things will have to change to accommodate a rotating Earth.

In light of the Aristotelian cosmological model, it is clear that the Earth does not move. This is the conclusion of the best science at the time. The stationary Earth, fit into the larger cosmology and physics, is exactly what follows from common sense and our mundane, untrained experiences. It is, in other words, exactly what you would believe if you hadn't learned otherwise in school. The stationary Earth is easy on the intuition, and it's wrapped up in sophisticated theory about space and matter and motion. It's part of a comprehensive and consistent worldview. That's hard to beat.

CHAPTER 4
Tinkering with the Standard Model

Certain people ... supposed the heavens to remain motionless, and the earth to revolve. ... But such things are utterly ridiculous merely to think of.
—**Ptolemy**

It would be misleading to say that the Aristotelian standard model presided continuously over the schools of cosmology from antiquity to the Renaissance without interruption or alteration. The doctrine was lost and found, at least by Europeans. Aristotle's texts and ideas disappeared from the Greek world but were transported to North Africa. They show up in Islamic philosophy and science where they were preserved and from which they were brought back to Europe in the late middle ages. From there we get transcriptions into Latin and introduction as the canon of medieval universities.

Along the way, alternatives were proposed, some that put the Earth in rotation. These generally appeared in isolation and failed to muster the rallying support of converts. No organized schools formed to weave new networks of ideas, no Pythagoreans or Peripatetics. There was no new evidence offered in support of any new model, no compelling empirical reason to put the Earth in motion. Nor was any clear and coherent network of theoretical ideas on motion or elements put in play, no system of physics to replace the Aristotelian worldview. Without this complete package of evidence and theory, there was nothing to seriously rival the standard model. We should nonetheless take account of the main alternatives to Aristotle that were on the books before Copernicus.

Within the Aristotelian school, details of the model were added and refined. There was some tinkering, but the fundamentals were unchanged. The center held for 2,000 years, and at the center was the stationary Earth. Lost then found, outlasting potential rivals, and with some revision, the Aristotelian standard model was the cosmology of record at the time of Copernicus.

Aristarchus of Samos is probably the most famous pre-Copernican to propose an outright alternative to the Aristotelian geocentric cosmology. Aristarchus lived from roughly 310 to 230 BC. He suggested that it is the Sun, not the Earth, that rests at the center of the universe, while the Earth, like the other five planets, orbits the Sun. The Earth also rotates on its own axis. It's easy to see why Aristarchus is often called the Copernicus of antiquity, and why he is a hero to modern Greeks. The airport on the Greek island of Samos is named for him. It's just a few kilometers from the town of Pythagorio, named for another famous proponent of the Earth in motion.

He is also a hero to many modern scientists. A NASA website refers to Aristarchus as "one of the greatest thinkers in human history." This is a conclusion drawn from very little evidence, really just one data-point. Only one written work by Aristarchus remains, and it's not about the location or motion of the Earth. In deciding on the location and movement of the Earth, we really don't know what he was thinking, other than that he came to the conclusion of a heliocentric universe. He may have been the Copernicus of antiquity, but with no information on how or why he got the idea, there is little to praise about his thought process. He may have been one of the greatest thinkers in human history, but we have no way to know this.

Usually on a science or math test you have to show your work to get credit. Aristarchus seems to have gotten a high score with the answer alone.

The one work by Aristarchus that does remain is on the relative sizes and distances of the Moon and Sun. It does not mention the center of the universe or the motions, relative or absolute, of the components. It neither describes nor presumes any cosmological model, and it works equally well with either the Sun or the Earth at the center, and with the Earth rotating or not. It's of no help to our cause, determining whether the Earth rotates or not. Nonetheless, it is an interesting and clever application of trigonometry. It depends on a precise measurement of a dodgy phenomenon, the position of the Moon when it is exactly at half phase. The Moon is then at the vertex of a right angle between the line from the Moon to the Sun and the line from the Moon to the Earth. The angular separation of the Sun and the Moon as measured from the Earth then reveals their relative distances from the Earth, and from this you get their relative sizes. Here we can follow Aristarchus' thinking, since he shows his work, and it is indeed praiseworthy. Unfortunately, his attention to precise measurement was less impressive. His report on the angular size of the Moon is off by a factor of four. As a result, his calculation on the actual distance to the Sun is too small by an order of magnitude.

The Aristarchus cosmology, a heliocentric model with a rotating Earth, is known to us primarily by reference in the work of his contemporary Archimedes. It's in the same treatise in which Archimedes estimated the number of grains of sands to fill the universe, the *Sand Reckoner*.

[Aristarchus'] hypotheses are that the fixed stars and the sun are stationary, but the earth is borne in a circular orbit around the sun, which lies in the middle of its orbit, and that the sphere of the fixed stars, having the same center as the sun, is so great in extent that the circle on which he supposes the earth to be borne has such a proportion to the distance of the fixed stars as the center of the sphere bears to its surface.

That's it. Archimedes offered neither praise nor endorsement of the hypotheses, and there is nothing about the derivation or reasoning or evidence. Archimedes, in other words, seems unimpressed, indicating that whatever evidence for the rotation of the Earth Aristarchus might have provided was neither noteworthy nor compelling. It's also unclear whether Aristarchus put the Earth in motion merely as one way to look at celestial phenomena, a perspective equally acceptable to the one with a central, stationary Earth. There's just not enough information to know whether the proposal was of a useful model or a real description of the universe. We just don't know what Aristarchus was thinking.

Copernicus credited Aristarchus, but it's unlikely he read about the Greek's idea in the *Sand Reckoner*. Lost and found, Archimedes' work did not reappear in Europe until its publication in 1544, 1 year after the death of Copernicus. Whatever the source, Copernicus made no reference to Aristarchus' reasoning, only his conclusion. There is the claim of a rotating Earth, but no evidence.

Outside of Europe, the movement of the Earth was considered and most often dismissed. The astronomical evidence, the detailed motions of the stars and planets, was frequently ruled to be equivocal, that is, in favor of neither the heliocentric nor geocentric system, neither a moving nor stationary Earth. Heavenly phenomena could be saved either way, with the the celestial sphere in orbit around a stationary Earth, or an Earth that rotates on its axis once a day. Deciding which model is true was left to theology or natural philosophy, that is, physics. Watching the stars and planets could only show a relative motion between heaven and Earth. Experiments on the Earth itself, and observation of natural terrestrial phenomena, were more informative. They were generally interpreted to show that the Earth does not rotate.

There were exceptions. Around 500 AD, the Indian astronomer Aryabhata included an enigmatic reference to the eastward revolutions of the Earth. It seems to mean the Earth's rotation. In context, the line is, "In a *yuga*, the revolutions of the Sun are 4,320,000, of the Moon 57,753,336, of the Earth eastward 1,582,237,500." The interpretation of this is challenging. A *yuga* is a Hindu expression of an era of human existence. The enormous numbers are explained in different ways by different Sanskrit scholars. And however the revolutions are described, there is nothing in the text to suggest the evidence or reasoning to support the conclusion. There is no evidence cited for the rotation of the Earth, if indeed that is what Aryabhata was describing.

The tenth-century Persian scholar Al-Biruni described an astronomical device that was apparently designed with the perspective of the Earth in motion. He reports

> I have seen the astrolabe called *Zuraqi* invented by Abu Sa'id Sijzi. I liked it very much and praised him a great deal, as it is based on the idea entertained by some to the effect that the motion we see is due to the Earth's movement and not to that of the sky.

The astrolabe was device for both measuring the inclination of a star or planet and doing some basic, slide-rule type calculations. Al-Biruni's praise seemed to be for the effectiveness of the tool rather than the cosmological insight. Astronomers accepted whatever worked to measure and forecast the positions of celestial objects, without regard for the truth of the matter. The Earth's movement could have been simply a working model, as effective as a stationary Earth but perhaps easier to build into the astrolabe. As to the reality of whether the Earth moves or not, Al-Biruni is among those who see the celestial evidence as incapable of deciding.

> For it is the same whether you take it that the Earth is in motion or the sky. For, in both cases, it does not affect the Astronomical Science. It is just for the physicist to see if it is possible to refute it.

Physics, recall, is the science of terrestrial phenomena.

In the thirteenth century, Nasir al-Din al Tusi, also Persian, had a similar assessment of the evidence. Astronomical data cannot show whether or not the Earth rotates. He was convinced the Earth does not move, but not on the basis of any empirical evidence. It's a matter of principle. Even Aristotle's demonstration of a vertically tossed stone falling straight back down does not prove anything. He explained

> It is not possible to attribute primary motion to the Earth. This is not, however, because of what has been maintained, namely that this would cause an object thrown up in the air not to fall to its original position but instead it would necessarily fall to the west of it, or that this would cause the motion of whatever leaves the [Earth], such as an arrow or a bird, in the direction of the [Earth's] motion to be slower, while in the direction opposite to be faster. For the part of the air adjacent to the [Earth] could conceivably conform to the Earth's motion along with whatever is joined to it …

With this analysis, Tusi acknowledged that what he took to be the true conclusion that the Earth does not move, could be reached on faulty evidence. In fact no evidence will suffice, and the stability of the Earth can only be known by principled reasoning. From the understanding of natural motion, the Earth has, "a principle of rectilinear inclination that it is precluded from moving naturally with a circular motion." With this conceptual foundation, the Earth's stability is immune from refutation. The immunity would be lost, however, on anyone who disagrees with the basic principle of natural motion.

The Aristotelian model of the cosmos was not perfect. Some of its shortcoming were recognized from the start; others emerged as astronomical data became more precise and more abundant. Ants on a potter's wheel will get you only so far in describing the heavens. Making the celestial bodies responsible for all of the observed motion, that is, leaving none of the task to a moving Earth, and requiring that there be nothing but perfect circles rotating with steady speed were challenging constraints. Three well-known celestial phenomena were not explained by the basic model. One was the retrograde motion. Planets wander steadily eastward along the ecliptic, until they stop, reverse course, stop again, and then get on with the trip east. Another problem was the observation that the planets gradually change in brightness, as if they are sometimes closer to us, sometimes farther away. There is a correlation between these two events, at least for three of the planets, Mars, Jupiter, and Saturn. The retrograde motion happens when the planet is at its brightest, and that is when they are opposite the Sun, high in the sky at midnight. The other two planets, Mercury and Venus, are never far away from the Sun, never even close to opposition. This is the third immediate failure with the basic Aristotelian model, as there is nothing about the ants on the wheel that explains the limited elongation of Venus and Mercury.

All three problems were fixed by a second-order application of Aristotelian principles. There were several contributors to the revision, and it took a few centuries. The best place to find the fully revised model is in the work of the Alexandrian astronomer Ptolemy. The *Megale Syntax*, Greek for the great compilation, is a monumental presentation of both observational data and theoretical modeling of the universe. Written in the second century AD, it was translated into Arabic in which the name became *Almagest*, simply the greatest. Ptolemy brought together contributions by earlier cosmologists, in particular Apollonius of Perga (third century BC) and Hipparchus of Rhodes (second century BC), added some of his own, and produced a system to predict the appearances of heavenly bodies forever into the future. It was a scheme to save the phenomena while staying faithful to Aristotelian principles.

Ptolemy kept the Earth stationary and in the middle of the universe. He added and refreshed the evidence for the stability of the Earth, but first let's get the basic structure of the revised model. The key was to get the ants out of their tracks. The planets and stars orbit the Earth westward in a 24-hour period—the spinning of the potter's wheel—and each planet is guided along on its own path to bring it very slowly eastward against the backdrop of the fixed stars. But a planet is not actually on the circle of its daily revolution; rather, it orbits a point on the circle as that point orbits the Earth. Thus, the so-called deferent circle is like the track on the potter's wheel around Earth, and the planet orbits on a smaller circle, an epicycle, that is centered on a guide point moving steadily around on the deferent. The trajectory of the planet is a rosette, a loop-de-loop pattern.

The full cosmological system with all the planets is a carousel of perfect circles and unchanging angular speeds.

This explains the changing brightness of the planets. As they circle around on their individual epicycles, they are sometimes closer to the Earth and other times further away. It also takes care of retrograde motion. As it orbits a point that is itself orbiting the Earth, a planet will at some times be moving backward, opposite to the direction of the guide point on the deferent. This happens when the planet is at its closest to the center of the deferent and closest to the Earth. The size and orbital period of each deferent and epicycle are adjusted to match the astronomical data, so that each planet's trip along the ecliptic and the timing of its retrograde are as predicted by the model. There are a lot of moving parts and a lot of adjustable parameters.

There was no third-order application of circular orbits, no epicycles on epicycles, but some adjustments at the center of the deferents were needed. None of these adjustments put the Earth in motion. That was unnecessary, since the diurnal phenomena, the daily risings and setting, presented no problems for the standard model. No rotation of the Earth was required to save the phenomena.

Planets wander in the fixed stars; that's why they are called planets. They travel eastward along the ecliptic, sometimes going backward, and they move at different rates in different parts of the sky. This last feature, taking longer to cross one part of the sky than the other, could not be accomplished with the basic deferent and epicycle system, let alone the ants on a potter's wheel. To account for the changing speed, the center of the deferent was moved away from the Earth to a point called the eccentric. The planetary epicycle provided the qualitative change of direction in a planet's movement—the retrograde motion—while the eccentric allowed for the quantitative change of speed and timing through the different stages of the trip around the ecliptic.

Even with the eccentric orbits, there was discrepancy between prediction and observation of how long it takes a particular planet to cover the distances through different parts of the ecliptic. Rather than compromise the Aristotelian principle of uniform rotational speed, that is, letting the guide point on the deferent simply speed up and slow down as indicated in the data, Ptolemy added one more detail. While the circular deferent is geometrically centered on the eccentric, the angular speed of the guiding point (the center of the epicycle) is determined from a different point. This additional point is called the equant. It is just as far away from the eccentric as is the Earth, but in the opposite direction. So, there are three points in a line: the equant, the eccentric, and the Earth. The deferent is centered on the eccentric, but it is the line between the equant and the center of the epicycle that sweeps around with constant angular speed.

This is faithful to Plato's challenge, to describe "the uniform and ordered movements by the assumption of which the apparent movements of the planets can be accounted for." It is often called the Aristotelian/Ptolemaic model, and it was the cosmological system taught in medieval universities and into the seventeenth century. This was what Copernicus had to work with. Strictly speaking you can't really call this model geocentric, since the Earth isn't exactly at the center. It goes by the designation, however, because the Earth is there roughly in the center, and it is the only object there in the middle. Whatever you call it, the Earth is not moving.

The clockwork of epicycles and deferents, offset by the eccentric and equant, may seem to us crazy and contrived. All of this, just to keep the Earth from moving. Ptolemy anticipated this reaction.

> Now let no one, considering the complicated nature of our devices, judge such hypotheses to be over-elaborated. ... Rather, we should not judge 'simplicity' in heavenly things from what appears to be simple on earth, especially when the same thing is not equally simple for all even here.

Deciding whether something is simple or complicated, in other words, is subjective. Our intuitions about how things go on Earth is a poor tutor on the nature of heavenly phenomena. Besides, most intuitions would agree that there is no simpler shape to model the repetitive motions of the stars and planets than the perfect circle.

There is no denying that the Ptolemaic system looks complicated to us. It seems like a desperate effort to protect the standard model from the threat of challenging evidence. It's surely not the only case in the history of science in which contrivances are added to plug the evidential holes in a favored theory. There is no better example than our own standard model of interactions among elementary particles, quantum electrodynamics (QED). The mathematics of QED present equations with solutions that are infinite, like dividing by zero. This can't represent a real, physical interaction, so a tricky mathematical procedure of subtracting out the infinites and leaving a meaningful residue was developed. It's called renormalization, and it has indeed become the new standard for hiding the troublesome explosions in the math. There is some discontent, and comparisons between renormalization and epicycles are not unknown. But for blunt criticism, trust Richard Feynman, Nobel Prize winning physicist and a key player in the development of QED.

> The shell game that we play ... is technically called 'renormalization'. But no matter how clever the word, it is still what I would call a dippy process! Having to resort to such hocus-pocus has prevented us from proving that the theory of quantum electrodynamics is mathematically self-consistent.

Hocus pocus or not, renormalization is a way to save both the phenomena and the core theory. We can say the same for epicycles.

It's possible that the Ptolemaic system was intended as simply a calculating device, a way of modeling the cosmos to facilitate prediction, with no regard for it being a true representation of the heavens. Ptolemy generally refers to his scheme as a hypothesis, suggesting that he was noncommittal as to the reality of the mechanism of deferents and epicycles. But it worked. It still does; modern planetariums project the images of stars and planets using a device designed with gears representing deferents and epicycles for the planets. The images move across the dome of the sky, as we sit comfortably still and watch.

Ptolemy may have been equivocal about the reality of the astronomical device, but he was clear in the claim that the Earth really does not rotate. He was equally clear in presenting the evidence. Like Aristotle, he explicitly considered the suggestion of rotation.

> Certain people ... supposed that the heavens to remain motion-
> less, and the earth to revolve from west to east about the same
> axis [as the heavens], making approximately one revolution
> each day.

The historical speculation is that he was referring to Heraclides. The rejection of the idea was somewhat less polite than Aristotle's, declaring that, "such a notion is quite ridiculous." Unlike Aristotle, Ptolemy did not find evidence in the sky to prove that the Earth stands still, admitting that, though the suggestion of rotation is ridiculous, "there is perhaps nothing in celestial phenomena which would count against that hypothesis." The Earth-bound evidence, on the other hand, he found to be decisive.

Ptolemy's proof that the Earth is not rotating was an elaboration of Aristotle's observations and conclusions about the trajectory of projectiles, objects tossed and allowed to fall to the ground. It's the explicit version of the intuition that drives modern-day deniers of rotation to point out that no one flies westward when they jump into the air, this despite the incredible speed of the ground on a rotating Earth. In Ptolemy's terms

> the revolving motion of the earth must be the most violent of all
> motions associated with it, seeing that it makes one revolution in
> such a short time; the result would be that all objects not actually
> standing on the earth would appear to have the same motion,
> opposite to that of the earth.

Even allowing that the atmosphere is somehow caused to follow the ground as it moves, solid objects such as our bodies and stones, compounds of earth and water, would not. These are the projectiles in the experiment. Ptolemy pointed out

> if they said that the air is carried around in the same direction
> and with the same speed as the earth, the compound objects in
> the air would none the less always seem to be left behind by the
> motion of both [earth and air].

And if these solids were in some way caught in the air and moved eastward with it, then they would be truly stuck and incapable of moving with any other horizontal speed. In his words,

> if those objects too were carried around, fused, as it were, to the air, then they would never appear to have any motion either in advance or rearwards: they would always appear still, neither wandering about nor changing position, whether they were flying or thrown objects.

None of these things are observed to happen. Dropped objects neither fall to the west nor remain frozen in the midair. And so, by the scientific logic of falsification, Ptolemy concluded that the Earth does not rotate.

Not everyone accepted this proof, including some who believed the conclusion and agreed that Earth is motionless. Medieval scholars began to question the Aristotelian physics of projectile motion. There had always been some discontent with the explanation of what keeps an object going after it is thrown and leaves the hand. Toss a stone straight up and natural motion should take over the moment you let go. Why doesn't the stone immediately drop straight down? What propels it to continue going up, albeit slowing down, to the top of the flight where the natural downward motion kicks in? Or toss the stone horizontally. Again, why doesn't it drop straight down when it leaves your hand? Why does it continue to move horizontally as it falls? The Aristotelian answer invokes a continuing violent force. As a stone or arrow flies off, it splits the air which must then rush around to the back of the projectile to fill the void. The air rushing in from behind pushes the stone or arrow forward.

The implausibility of this explanation had implications for arguments about the rotation of the Earth. There was no problem with celestial motion, as long as it was composed of perfect circles and uniform speeds. Such purely natural motion requires no cause; left alone it will continue eternally. But terrestrial proofs of the stability of the Earth involved violent motion of projectiles and required an accurate understanding of the cause.

Jean Buridan, a fourteenth-century French philosopher and scientist, proposed an alternative explanation of the motion of a projectile. He introduced the idea of impetus, a residue of force left in the stone, imparted by the throw.

> [The projector] impresses a certain impetus or motive force into the moving body, which impetus acts in the direction toward which the mover moved the moving body, either up or down, or laterally or circularly.

This sounds like inertia, but it's fundamentally different. Impetus is an active force, implanted in the projectile by the projector, the bow to the arrow, or the hand to the stone, and continuing at work through the flight of the projectile. Inertia, our modern conception, is the continuing motion in complete absence of any force. Impetus is the leftover force causing some dwindling violent motion; inertia is natural motion. Throwing a stone is an act of violent motion, and with impetus, the violence continues even after you let go. In Buridan's account, the natural motion of a stone is straight down, just as in Aristotle's physics.

Nicole Oresme was a student of Buridan. He made no explicit reference to impetus, but the concept is clearly at work in his analysis of the evidence for the stability of the Earth. The argument of projectiles, tossing a stone or shooting an arrow straight up, proves nothing, according to Oresme. Of course the object falls straight back down if the Earth is stationary, but it would also fall straight back down even if the Earth was rotating. On a rotating Earth

> an arrow shot straight into the air is [also] moved rapidly eastward with the air through which it passes and with the whole mass of the bottommost [or terrestrial] portions of the universe described above, the whole [earth and air and arrow] being moved with daily rotation. Therefore the arrow returns to the spot on the earth from which it was shot.

Arrows, stones, clouds, and a person jumping would all receive the horizontal impetus from the hypothetically rotating Earth.

Consequently, no experiment on the Earth itself, and no natural terrestrial phenomenon would turn out any differently whether the Earth rotates or not. Stuck as we are on the ship, there is nothing we can observe on board to demonstrate our state of motion.

There is nothing we can see looking out that will do it either. Oresme, in agreement with Ptolemy, argued that the celestial evidence can always be explained either way, on a stationary Earth or on an Earth in rotation. Observation can only show relative motion, that one body is moving relative to another. In Oresme's words, "I suppose that local motion can be perceived only when one body alters its position relative to another." There can be no direct observation that something is moving absolutely.

That doesn't mean there can't be indirect evidence of absolute motion, and it's useful to distinguish between what is sometimes called optical relativity and dynamic relativity. The former is about what can, or cannot, be seen. You could also call it kinematic relativity, since kinematics is the physics of describing motion, how things move. Dynamic relativity is about what can, or cannot, be explained. Dynamics is the physics of the causes of motion, that is, why things move. There may be things that can only be explained in terms of absolute motion. If so, then we may not be able to see it directly, but we know it's happening because we see the effects. Aristotle and Ptolemy claimed that the straight vertical fall of a stone could be explained only by a stationary Earth; this implies that the Earth is absolutely at rest. Oresme, with an alternative dynamic theory of projectile motion, one that invoked impetus, countered that the straight vertical fall could be explained either way, by rotation or stability of the Earth. This implies that, as far as the evidence showed, both celestial and terrestrial evidence, it is impossible to tell whether the Earth rotates or not.

Oresme nonetheless believed that the Earth does not rotate. There may be no perception or even indirect evidence of absolute motion or rest, but he maintained it is the fact of the matter nonetheless. It is meaningful to say the Earth rotates without reference to any other objects. It's meaningful but, according to Oresme, it's false. He did not give any clear reasoning to support his conclusion that the Earth stands absolutely still. And, of course, he provided no evidence, since, by his arguments for both kinematic and dynamic relativity, no evidence is available. This missing support for the conclusion about rotation may explain why neither Copernicus nor Galileo said a word about Oresme.

Oresme's work helps in highlighting an important distinction in the evidence for the rotation of the the Earth, the two perspectives, celestial and terrestrial. One is the evidence in the sky, the other on the ground. We've seen arguments that one or the other, or both, cannot provide evidence for or against the rotation of the Earth. Ptolemy admitted that astronomy, the source of celestial evidence, could decide nothing on the motion, or not, of the Earth. Oresme suggested that the terrestrial evidence was equally uninformative. Later and in more detail, Galileo will argue that no experiment on . the Earth itself can reveal whether we are moving, agreeing that the terrestrial perspective is blind to rotation. He will do this by revising the physics of projectiles, from impetus to inertia, keeping the on-going horizontal motion of any projectile such that it always follows along with a rotating Earth. He will invoke, in other words, the dynamic relativity. It becomes known as Galilean relativity.

If neither perspective can provide evidence for or against rotation, as Oresme argued, we would be left in a position of indecision. Cosmology presents two hypothesis, soon to be described by Galileo as the two chief world systems, that seem to have equal evidence and no available data to decide which is true. It's a situation for Buridan's ass, an allegorical beast satirically named for a philosophical conundrum described by Oresme's teacher. A hungry donkey, left standing at equal distances from two temping servings of hay will starve to death, paralyzed by indecision. Buridan's discussion was about human determinism, making a case against the possibility of freewill. But the analogy is food for thought in understanding the evidence about rotation.

CHAPTER 5
Moving the Earth

... to ascribe movement to the Earth must indeed seem an absurd
performance on my part ...
—**Nicolaus Copernicus**

As individuals, our knowledge that the Earth rotates is not based on personal observation or any experiment we can do ourselves. In fact, all our own experiences tell against it. We almost surely get the idea from a teacher or by reading about it in books or on the Internet. The knowledge is from authority, properly chosen. On matters scientific, you want to learn from a scientist, not a movie star or a famous athlete or some crackpot blogger. If it's in a book you check the publisher, looking for a reputable and careful production of textbooks or monographs. Or if you're on the web, choose a .edu over a .com, at least as a general rule. There are always exceptions.

Science always entails some surrender to authority. It also demands some tough-minded independence and skepticism. There is this inescapable balance to achieve, weighted by two contradicting influences. You want to benefit from the expertise of others but also think for yourself. If one or the other of these dominates, not much science gets done. Leave each of us to our own thoughts and experiences, and our individual knowledge of nature will be naïve, narrow, and ignorant (perhaps blissfully) of other perspectives. And, of course, ignoring the scientific conclusions already on the books will make the process of learning about nature wildly inefficient. But too much trust in the authorities prevents significant challenges to what is confidently in the textbooks, allowing misconceptions to remain and preventing the sort of change we count as scientific progress. It would certainly impede anything that would amount to a scientific revolution. The tension between authority and challenge, between the inertia of established ideas and the disruption by novelty, is what holds the scientific method together. It's an essential part of the process.

Publication of scientific ideas is an important influence on both sides of the balance, giving weight to the established account of nature and voice to the challenges. The eponymous Nicolaus Copernicus developed and published his cosmological model in the sixteenth century, just after the invention of the movable-type printing press. Books facilitated access to the authoritative word of the standard model and Aristotelian physics, as textbooks now give every student a detailed picture of the modern scientific description of nature. Printing was not only faster than transcription by hand, it was more reliable. Each copy of a book would be the same, avoiding the idiosyncratic errors that were all but inevitable when each manuscript was copied by hand. The contents of a book may be false, but at least there was some reassurance that the message of the author was faithfully reported. The standard model was effectively standardized.

Books gave their printed contents an authoritative presentation. Science could now be written down and widely shared, and scientists could do their work in the library rather than the laboratory or observatory. With so much knowledge there on the page, it seemed possible to learn about nature without the bother of making one's own observations or measurements. The term for this is "scholasticism", and it's not meant to be praise. Studying the scholars as a way of doing science makes challenges unlikely and makes it all but inevitable that mistakes will be faithfully repeated. The printing press encouraged scholasticism, and this weighed on the authoritative side of the scientific balance.

But don't blame the technology for perpetuating inaccuracies or, in particular, the endurance of the Aristotelian/Ptolemaic standard model. The model held authority for nearly two millennia before the invention of the printing press, and it began to crumble just a century after. New ideas generally come from old ideas, by revision or by explicit rejection. The publication of both the corpus of Aristotelian cosmology and physics, and various challenges to the establishment, weighed in on both sides of the balance between authority and novelty. Books were an important catalyst in the scientific revolution. The rapid and reliable reproduction of information made alternative cosmological options available. A variety of competing worlds on paper could be studied, and in this way the technology of printing could work to undermine rather than enforce authority. Books offered new ideas and challenged the weaknesses in the established theories. Books, like the Internet, provided a garden of ideas for the reader to look over, but there is always weeding to be done. Any novel suggestions would have to be tested, answering to the empirical data, saving the phenomena.

There were some flaws in the Aristotelian standard model, and now you could read about them. These were generally regarded as problems within the system, details to be fixed in order to maintain the basic cosmology. One was about the heavenly spheres themselves. In the original first-approximation by Aristotle, these spheres were made of quintessence, solid and crystalline and sturdy enough to hold the stars and planets. It was the crystalline spheres themselves that revolved around the Earth. But with the addition of epicycles, the mechanics of revolutions within revolutions seemed to involved spheres crossing paths and moving through each other— an impossibility for solids. This problem goes away if the Ptolemaic model is unreal, just a mathematical calculating device to save the phenomena. But if you wanted to know what's really out there, the intricate mechanism of heavenly spheres needed to be explained, or at least clarified.

There was also some concern about comets, so-called bearded stars. These were thought to be below the Moon and within the upper level of the Earth's atmosphere. They couldn't be celestial objects, beyond the Moon and heavenly, since they are ephemeral, coming and going, unlike the eternal and unchanging nature of the heavens. Aristotle dealt with comets in a book on meteorology, not in the book on astronomy. But comets clearly move horizontally, and they just as clearly have no influences forcing them from what should be their natural motion. As terrestrial objects, their natural motion is straight up or down. The explanation, available to be read in books, was that apparently the highest levels of the atmosphere are dragged around by the circular motion of the lunar sphere that rubs up against it. The air up there must be circling the Earth, a motion forced by the sphere of the Moon that drags the comets along with it. Not everyone found this explanation satisfying.

We have seen that the term "geocentric" is a bit misleading when applied to the Ptolemaic model of the universe, because the Earth is not really at the center of the things. Deferents are offset by the eccentric, and worse, the measure of uniformity of circular motion is moved to the equant. To some medieval astronomers, the equant in particular violated the Aristotelian celestial code for building the model out of perfect circles moving around with unfaltering constant speed. There was theoretical work still to be done, work to restore the ideals of circularity and uniformity.

Just as some of the conceptual details of the standard model were reportedly imperfect, so were there empirical difficulties. Predictions of astronomical events and positions of planets and stars were failing to match the observations. The phenomena were no longer being saved, at least not exactly. But the Ptolemaic model had a lot of adjustable parameters, the sizes and periods of deferents and epicycles, and it was generally assumed that with some appropriate adjustments the model could be made to fit the data. It wasn't so much the evidence that was challenging to the standard model, as it was the publication of its own internal conceptual failure of not measuring up to Aristotelian standards.

Note that none of these problems with the Ptolemaic cosmology directly involve the rotation of the Earth. They are all about orbits of heavenly spheres, and the expectation was that they would be solved by refinements in astronomy.

It was the equant that most bothered Copernicus, the grit in the oyster that would result in a pearl. Georg Rheticus, Copernicus' protégé who would ultimately see the Copernican *On the Revolutions of the Heavenly Spheres* through the first stages of publication, called the equant "a relation that nature abhors." With an equant embedded near the center, the Ptolemaic model was not sufficiently Aristotelian. Copernicus was the cosmologist of record for the revolution, but in many ways he was one of the last of the ancients, working to bring cosmology back to its roots by making it more faithfully Aristotelian. Tycho Brahe, often regarded as the most accomplished astronomer before the use of telescopes and a dedicated believer in Aristotelian principles, called Copernicus "a second Ptolemy."

Nicolaus Copernicus was born in 1473, making him 19 years old when Columbus sailed across the Atlantic. It was a time of discovery and there were changes being made in the understanding of the world. His family in Torun, recently under Polish authority, was financially comfortable, but his father died when Nicolaus was 10. An uncle assumed the role to oversee Copernicus' education. He was enrolled at Cracow University when he turned 18, an institution regarded as a center for astronomy and astrology. He left after 4 years, without a degree, but apparently with an interest in astronomy. There is no way to know if that's where he acquired his distaste for the equant.

Copernicus then went to Italy, to the University of Bologna, to study the canon law of the Catholic church. The plan seemed to be that he would follow the career path of his uncle, a Catholic bishop. Bologna offered an advantage over Cracow by having Greek in the curriculum, and learning that language gave Copernicus access to the Greek astronomers. And since he lived with a professor of astronomy, the celestial interest was cultivated and he began to make a few observations himself. His connection to the Catholic administration, though, continued, and while still a student in Bologna, Copernicus was elected canon of Frombork cathedral, on the Polish coast of the Baltic sea. Not quite a priest, he was on the committee that oversaw the administration of the cathedral. There were advantages to this, and even when absent from the city of Frombork he was provided with a stable income.

Copernicus left Bologna without earning a degree from the university. He moved to the University of Padua, this time to study medicine. There was some link between medicine and astronomy at the time, in the discipline of medical astrology. Since the arrangement of celestial bodies was thought to have some influence over the creation of a human body and spirit, a knowledge of the stars and planets was beneficial to the physician. It is another matter whether knowledge of medicine is of value to someone whose real interest is astronomy, and Copernicus left Padua without a degree. Finally at the University of Ferrara, he received a degree in canon law, as he had begun his studies in Bologna.

Returning to Frombork, Copernicus built himself a small observatory. The edge of the Baltic sea is not an ideal setting for astronomical work, given the low elevation and frequency of fog, and he did not add significantly to the record of celestial evidence. He seemed content to accept data already on the books. Sometime around 1514, he wrote and distributed a small treatise on the general structure of the universe. The *Commentariolus*, the *Little Commentary*, outlined the basic ideas of a heliocentric universe in which the Earth is in motion. It was never published, only shared with a few friends and colleagues in hand-written form. The *Little Commentary* sketched the theory that the Earth moves, but it did not include any new evidence or any novel reply to the arguments by Aristotelians citing the terrestrial phenomena that demonstrate the stability of the Earth. There was no explanation of the author's formative reasoning, nor an attempt to convince a skeptic. It was just the beginning, and there was more work to be done.

What began in the *Little Commentary* was completed by 1543 with the publication of the full Copernican model in *On the Revolutions of the Heavenly Spheres*. It took a long time to do the math and attend to the details of the system, putting the planets in order and putting the Earth into motion. And it was a long road to publication, both figuratively and literally. With no nearby publishing house, the manuscript was taken to Nuremberg, Germany for the final printing.

The finished work made it back to Copernicus in Frombork just before his death.

The basic heliocentric model of the cosmos is presented in Book One of *On the Revolutions*. It worked perfectly as an answer to Plato's challenge: "What are the uniform and ordered movements by the assumption of which the apparent movements of the planets can be accounted for?" Anything that moves in the heavens does so on a perfect circle with an eternally uniform speed. Since this is the natural motion of celestial objects, there is no concern for a cause to either start or maintain the rotations and revolutions. The crystalline spheres of the Aristotelian/Ptolemaic system are also in the Copernican model; these are the heavenly spheres named in the title. All of the fundamental structural features are Aristotelian, and the phenomena are saved without the offending equant.

The features of the universe in Book One of *On the Revolutions* are pretty much what we now learn in our first science class in school. The Earth rotates, once around every 24 hours. This explains the day and night. It also revolves around the Sun, once around every year. This explains why the Sun appears at different places against the pattern of fixed stars, its annual trip along the ecliptic, because the pattern of stars is genuinely fixed. The Earth is just one of the six planets, all in solar orbit, with different periods, correlated to their different distances from the center. This explains each planet's movement along the ecliptic, including the retrograde motion. When the Earth's orbit overtakes that of another planet, that planet appears to be moving backward.

From a modern perspective, from this side of the revolution, some of the Copernican explanations seem more natural than what was offered by Ptolemy. The Ptolemaic model can accommodate just about any celestial phenomenon, but it may be by an *ad hoc* adjustment of some of the many variable parameters. It has a lot of moving parts, and so it is amenable to a lot of tinkering. The Copernican model turns out to require a significant number of pieces as well, but there is linkage between the parts that results in an admirable harmony of movement and coherence of explanation. The correlation between a planet's orbital period and its distance from the Sun, for example, means that from the measurement of how long it takes to complete a circuit of the ecliptic you then know the planet's relative distance. One parameter is fixed by the other. And the curious limitation on the elongation of Mercury and Venus, their angular separation from the Sun, follows inevitably from their being in orbit around the Sun, both closer than the Earth to the center. With these connections, the new system is more coherent, revealing the orderly structure we might expect to find in nature.

The heliocentric model thus gracefully solved some problems for which the geocentric system could only make clumsy excuses. But the new model brought on problems of its own. Changing the position from which you view a stationary object will change the angular direction of the image. It's called parallax. Blink from eye to eye and note how a nearby object, your thumb at arm's length, for example, appears to shift back and forth. If the Earth is in motion while the stars are stationary, the direction to any given star should appear to shift over time. This would happen most notably over a 6-month interval, as the Earth travels from one side of the Sun to the other. This should be a way to test the Copernican heliocentric hypothesis, by choosing an identifiable star and measuring its position to detect the parallax. It's a test the hypothesis failed.

Parallax depends on distance; the farther away the object, the less it will appear to shift. Blinking eye to eye you can see the parallax of your thumb, but try it on something across the room or down the street and the effect will be undetectable. It's there, just too small to see. This is the way to reconcile the heliocentric model with no evidence of parallax. The effect is there, but just too small to detect, because the stars are too far away. It's not that this celestial evidence is equivocal in principle, as many of the medieval astronomers argued, but it is uninformative in practice. Both theories, with the Earth in motion or not, were compatible with the data as far as one could tell.

If there is no way to measure parallax, it being too small an effect, and hence no way to tell one way or the other if the Earth orbits the Sun, then perhaps it doesn't matter which model we choose to work with. Either perspective works to save the phenomena, so opt for the tidier, cleaner calculations. On this interpretation, think of Copernicus as saying simply that the universe is *as if* the Sun is at the center and the Earth has two motions, rather than the Sun *really is* at the center and the Earth really moves.

Exactly this interpretation was advocated by Andreas Osiander in an unauthorized and anonymous preface to *On the Revolutions.* Tasked with seeing the book through its final steps of publication, Osiander apparently thought to shield Copernicus from disapproval by the Catholic church and the scriptural declaration that the Earth stands still. There is no outright claim that the Earth is in motion, Osiander prefaced, "for it is not necessary that these hypotheses should be true, or even probably; but it is enough if they provide a calculus which fits the observations." Copernicus never approved this anti-realist status of his cosmology, and the details of his reaction to reading Osiander's interpretation are unclear. Some of his followers, though, left nothing vague about their disapproval. Johannes Kepler described the preface as "written by a jackass for the use of other jackasses." Giordano Bruno called the (still anonymous) author of the preface an "ignorant and presumptuous ass."

At the time of Copernicus, the celestial evidence may have been compatible with either model of the universe, but what about terrestrial phenomena? Ptolemy had argued that when you look carefully at things that happen on the Earth itself, the tossing of stones and the shooting of arrows, for example, the hypothesis of a rotating Earth was shown to be false, or, as he put it, utterly ridiculous to think of. Copernicus confronted this evidence directly. His efforts were not to prove that the Earth rotates, only to refute the previous arguments that allegedly showed the Earth does not rotate. The evidence in what we observe on the Earth is entirely compatible with the Earth in rotation. It's a possibility, a real possibility.

The Copernican case for rotation began with a noncommittal, suggestive tone.

> And since it is the heavens which contain and embrace all things in the place common to the universe, it will not be clear at once

why movement should not be assigned to the contained rather than to the container.

In other words, it could happen; maybe the celestial sphere of stars remains in place while the Earth rotates.

Ptolemy had insisted that the incredible speed of a rotating Earth would cause unsecured objects to fly off. Even pieces of the Earth itself would be thrown free, and our solid home would break apart, the bits dispersed into the heavens. Copernicus replied by pointing out that on the Aristotelian/Ptolemaic model, the celestial sphere of fixed stars rotates much faster, it being so much larger yet still getting all the way around in one day. Ptolemy's logic leads to the conclusion that the whole spinning cosmos would fly apart. In this light, the simple rotating Earth looks more plausible.

But there was the frequently raised objection that on a rotating Earth, loose objects left to their natural motion would be left behind. This is not just about dropped stones and shot arrows, but even the clouds and air would be seen to rush to the west if the Earth rotates eastward. Here Copernicus offered two possibilities that would keep the clouds in place on the moving Earth. It is

> because the neighboring air, which is mixed with earthly and watery matters, obeys the same nature as the Earth or because the movement of the air is an acquired one, in which it participates without resistance on account of the contiguity and perpetual rotation of the Earth.

The air and clouds, in other words, are dragged along by their contact with the ground. As for the dropped stone

> we must confess that in comparison with the world the movement of falling and of rising bodies is twofold and is in general compounded of the rectilinear and the circular.

Because it has been lifted off the ground, the stone will have a natural motion straight down. It's worth pointing out that the reason the stone falls is not as clear in the Copernican system as it was with Aristotle. With the center of the universe no long coincident with the center of the Earth, the natural place of earth (the element) is no longer the natural center of the Earth (the planet). There is no explanation of the vertical fall of a dropped stone. Nonetheless, Copernicus retained this detail of physics, the rectilinear movement, and he added to it. Because the stone is a piece of the Earth, it will share the natural circular motion of the sphere. As the air is dragged around by proximity to the ground, so too is a solid, earthen, projectile.

None of this is evidence that the Earth rotates. It only shows that what is observed in terrestrial phenomena is compatible with rotation. The same is true of comets, still thought to be at the interface between heaven and Earth. Comets were an awkward fit in the Aristotelian/Ptolemaic system and required a blurring of the celestial/terrestrial distinction. The upper atmosphere, where comets appear, must somehow acquire some of the motion of the contiguous lunar sphere in its diurnal rotation around the Earth. The "somehow" is telling, and Copernicus saw an opportunity to exploit the uncertainty. If the upper atmosphere could be dragged around by what rotates above it, then the lower atmosphere, the air we feel and the clouds we watch, could be dragged around by what rotates below it. Once again, the evidence could go either way.

Copernicus' efforts showed that the evidence, both celestial and terrestrial, is equally consistent with either conclusion, the Earth does or does not rotate. This at least opened the possibility for rotation. We see the Sun rise and set, and the firmament drifts across the sky during the night, but this could be just the result of our own moving perspective. He quoted Virgil, "And things are as when Aeneas said in Virgil: 'We sail out of the harbor, and the land and the cities move away.'"

No single observation shows the Earth to be rotating, but when you put all the evidence together, including the coherent mechanism of a heliocentric solar system, Copernicus concluded that it is more likely than not that the Earth rotates. "You see therefore that for all these reasons it is more probably that the Earth moves than that it is at rest." He offered no explanation why it moves, and no account of the sustaining cause of the motion, other than to say that it is natural for a sphere to rotate. This is not an unusual default explanation in science; it's just natural.

The systemic coherence and harmony of the heliocentric model carried a lot of the burden for proving the individual component that the Earth rotates. But the model as we've described it so far, and as Copernicus laid out in Book One of *On the Revolutions*, is less accurate than the Ptolemaic model. Using the basic heliocentric model to calculate positions of planets suffers larger errors than result from the Ptolemaic geocentric system with its deferents and epicycles. You have to look deeper into the book to find the details that bring the Copernican model into comparable compliance with the data. These are details left out of our science classes.

To account for the annual seasons and the Sun's changing position in the sky, Copernicus added a third motion to the Earth, an annual precession, a wobble of the axis of rotation. The Earth revolves around the Sun and rotates on its own axis. For reasons still unknown, the axes of the two motions are not aligned; the rotation axis tilts at about 23°. When the north pole is tipped out away from the Sun, the season is winter in the north and summer in the south. In order to change the season, Copernicus had the rotation axis precess so that in 6 months the south pole would tip away. In fact, this means that the rotation axis maintains a constant alignment with respect to the fixed stars, and the seasonal change is a result of the annual orbit alone. Whether the precession is a third motion or not depends on the choice of reference frame. Copernicus counted it as a third motion for the Earth, and that matters when one appeals to the simplicity of the model as a factor in deciding to accept it. In Copernicus' own view, the change from the Aristotelian/Ptolemaic to Copernican system increases the number of motions of the Earth from zero to three.

More important to the accounting of complexity are the epicycles. The Ptolemaic universe is rich with epicycles, but so is the Copernican. Deeper into *On the Revolutions*, each planet is put on a small epicycle. They are small enough, and the period of the epicycle orbit is slow enough, that the planet does not cross its own path to create the loop-de-loops that Ptolemy used to create retrograde motion. The effect of the smaller Copernican epicycles is to subtly elongate the planet's path around the Sun such that the trajectory is an oval. To accomplish this, the Copernican model ends up with roughly as many epicycles as the Ptolemaic. The result is a match with the astronomical data that is also comparable.

With the eccentric and equant, the geocentric Aristotelian/Ptolemaic model was not really Earth centered. With some complication in the middle, it turns out the heliocentric Copernican model is not really Sun centered. Again, beyond Book One, things get messy. The Sun is indeed stationary. In this way, there is a straightforward Earth-Sun switch between the two models. In one system the Earth is absolutely at rest; in the other the Sun is absolutely at rest. But in the Copernican system, the Sun is not the center of the Earth's orbit, nor the orbit of any of the other planets. In fact, the center of the planetary orbits moves. It orbits on a small epicycle on a deferent that is centered on the Sun. In other words, the Earth orbits a point that orbits a point that orbits the Sun. This accounts for the changing rate of the Sun's passage along the ecliptic and brings the model into better agreement with observations, matching the empirical success of Ptolemy.

None of these extra details, the epicycles and the off-center orbit, directly affect the rotation of the Earth. Copernicus had provided some speculative arguments to make the terrestrial observations compatible with rotation, but that was not evidence for rotation, particularly not in light of the most basic sensations of an unmoving Earth. With the publication of *On the Revolutions* there was no specific evidence of rotation, and certainly no direct observation of the phenomenon itself. The reason to believe the Earth rotates was that rotation fits into the larger cosmological system that saved the astronomical phenomena. Each component of the model is believable because the whole system has both empirical and conceptual credibility. That makes the details of the system, including the small complications, relevant. The Book-One basics, and what we get in school, are compelling in their simplicity and tidy coherence. The natural order of the planets, the appearance of their retrograde motion, and the explanation of the limited elongation of the inferior planets come together to make the Copernican model seem right and to motivate a Copernican to suspend the personal sense of stability of the Earth. But the deeper details—the small planetary epicycles, the complexity at the center, and the lost explanation for why things fall to the ground—were reasons to hesitate in adopting the new world system. A case was made for the Earth rotating, but it was not without some reasonable doubt.

CHAPTER 6
The Best of Both Worlds

... the earth, that hulking, lazy body, unfit for motion ...
—Tycho Brahe

Day to day, science generally progresses by the accumulation of data. There is the expectation of higher resolution in the picture of nature, and a sharper focus on the truth, with more points of light. And it's not just more evidence you want, it's more precision in the measurements and more reliability in the information. Forecasting the weather, for example, and understanding the dynamics of climate change, require the readings and records of instruments in as many places as you can afford and manage. Atmospheric models are empty and idle without some specifics of initial conditions to put the pieces in place. Research money usually goes to the lab or the observatory to provide the details that refine and test the theories. Big science often devotes big resources to finding just a few crucial parameters to anchor a theory. The large hadron collider and the search for the Higgs boson are the clearest examples. But the science never stops, and there are always more studies to fund, delivering more empirical details. Science is, as scientists say, data driven.

Copernicus did almost none of this. Neither did Einstein. Copernicus accepted the tables of astronomical observations already on the books, accumulated since Ptolemy without significant improvement in precision. His idea of scientific progress was not to add new pieces to the puzzle but to rearrange the pieces already there, abiding by the established laws of nature to control how things fit together. He ended up with a new picture of the cosmos, but with no better match to the evidence.

The evidence to guide and constrain cosmological theories was no better in the mid-sixteenth century than it was in the second, when Ptolemy produced the *Almagest*. This includes evidence on the rotation of the Earth. Data in astronomy and physics usually boil down to measuring positions, where things are and when they are there. In astronomy, it's the exact positions of planets and stars. This means their angular positions relative to each other or to the horizon. With astrolabe and cross-staff, late-antiquity and medieval observers achieved a precision of roughly a sixth of a degree, 10 minutes of arc. That's about a third the diameter of the Moon.

Compare this precision in the instrumentation with the estimated size of the effects that would put a world system to a test. To know if the rotation of the Earth is detectable, for example, the speed of rotation must be stated. That requires an estimate for the size of the Earth. The radius of the spherical Earth was measured by Eratosthenes in the mid-third century BC by comparing the angle at which sunlight hits the ground at two widely separated places. His value of 6,700 km was corroborated and pretty much unchanged through the time of Ptolemy and Copernicus, and it's remarkably close to the 6,400 km as measured today. Spinning once around in 24 hours, the ground at the equator will be moving at 1,700 km/hr. At a latitude of 45° it's a little slower but still a brisk 1,200 km/hr. That's plenty fast to detect a free-falling object being left behind, even in the short duration of air-time when dropped from a modest height. If that's the test for rotation, there is no excuse that the effect is too small to measure.

The other possible motion of the Earth, the orbit around the Sun, is a different matter. Parallax, the apparent change of position of an object when viewed from different positions, depends on both the distance to the object and the separation between the two points of observation. If the effect of stellar parallax, brought about by the Earth's annual trip around the Sun, is anything less than the instrumental limit of 10 minutes of arc, it will be unobservable. The only way to know is with values for the distance to the stars and to the Sun, and a little bit of trigonometry.

Ptolemy had a way to calculate the relevant parameters. He started with the Moon and worked his way out. The distance to the Moon he got by parallax measurements from two places on the Earth, an effect that would be the same whether the Earth moves or not. The radius of the Moon's orbit varies a little bit, but Ptolemy determined the maximum to be 64 times the radius of the Earth, 64 R_E. This sets the minimum for the orbit of the next celestial object, Mercury. All the planets in the Ptolemaic system orbit the Earth on epicycles and deferents, coming closet to us when the epicycle spoke points in, cycling farthest away when it points out, toward the celestial sphere of stars. Each planet orbits around in a band, the width prescribed by the diameter of the epicycle. And length of each epicycle radius can be determined from the details of the planet's movement along the ecliptic. There is no empty space between the heavenly spheres that carry the planets, so Mercury's sphere brushes against the lunar sphere below and that of Venus above. This means that Mercury's distance from the Earth varies from the minimum 64 R_E to a maximum that adds double its epicycle radius. Mercury's maximum sets the minimum for Venus, and so on for the rest of the planets. Ptolemy put in the parameters of planetary orbits, including the radius of each epicycle, and determined that Saturn, the most distant planet, turns out to be almost 20,000 R_E from the Earth. And that must be the distance to the stars, since they are all embedded in the single outermost celestial sphere that fits with no gap at the top of Saturn's orbit.

It's interesting to note that this astronomical technique of using what you can measure in nearby objects and then extending the information to the more distant is still common practice. Ptolemy measured the distance to the Moon and used information on epicycles and deferents to get stellar distance in terms of lunar distance. Now astronomers measure the distance to nearby stars by parallax.

They then use characteristic properties of particular kinds of stars, for example, the spectrum of electromagnetic radiation or the period of variable brightness, to figure out the absolute luminosity of these stars. Distant stars of the same kind are assumed to have similar correlations between luminosity and these sorts of properties, and so their absolute luminosity can be known. Comparing this to their apparent brightness gives the distance to stars that are too far away to detect parallax.

The Ptolemaic world system put the stars and the edge of the universe at 20,000 R_E away, 130 million km. That's a lot smaller than what we now measure, with the closest stars about four light-years (4×10^{13} km) away, and the most distant galaxies in the range of billions of light-years. But in the context of the time this was truly astronomical, the hugeness of the heavens to humble life on Earth. And consider the speed the stars have to be going in a system of this size in which the celestial sphere orbits the Earth in 24 hours. Its 34,000,000 km/hr. Looking out at the peaceful night sky, it's hard to imagine the stars scattered along the ecliptic being that far away and moving that fast. But again, science makes us realize that there are lots of things about the universe that are hard to imagine.

Along the way from the Moon to the edge of the universe, Ptolemy found the distance between the Earth and the Sun to be 1,200 R_E. That's 8 million km, compared to the current value of 150 million km. What counts in assessing the possibility of measuring stellar parallax and the resulting evidence that the Earth moves is the ratio between the distances to the stars and to the Sun, 20,000 R_E and 1,200 R_E, respectively. The angle between two sides of an isosceles triangle that are 20,000 units long where the opposite side is 1,200 units is 10°. That would be easily detectable at the time of Ptolemy or Copernicus. Given the best empirical data at the time, the sizes of astronomical orbits and the results of looking for parallax with the most refined and reliable instruments, a prediction of the Copernican model that should have been clearly observed was not. The excuse of the Copernicans was that the distance to the stars must be somewhat larger than Ptolemy reckoned.

It's fair to say that with the publication of the new world system, the heliocentric Copernican model, there was no overwhelming evidence in its favor. Neither world system enjoyed an empirical advantage, and resistance by either side was neither dogmatic nor denial of the obvious. There was reason for honest scientific disagreement. From our perspective, the heliocentric system brought conceptual and esthetic improvements, a harmony and coherence missing from the geocentric contrivance of deferents and sweeping epicycles. That's the room for disagreement, conceptual, and esthetic.

Both systems suffered some small empirical inaccuracy, missing predictions of planetary positions, and other important astronomical and astrological events. Copernicus tried to improve the match between theory and evidence by changing the theory. There is another way to do it, by refining the measurements that set the internal parameters of the existing theory; keep closer track of the planets in order to get the exact sizes of deferents and epicycles. This was the approach taken by Tycho Brahe.

Born into Danish nobility, Tycho, as he is usually called by historians, was only 17 when his interest in astronomy was confirmed by a predicted conjunction of Jupiter and Saturn. This was August of 1563. The dramatic meeting of the two bright planets occurred, but nearly a month later than had been forecast by the Ptolemaic model. Using the Copernican system was a bit better, but still off by days. It wasn't enough to convert Tycho to the new heliocentric description of the universe, but it did convince him of the sorry state of astronomical prediction and the need for greater precision in the instruments for observing the heavens. That became his occupation.

Tycho, like Copernicus, was raised and mentored by an uncle. He also spent an itinerate youth, moving from one university to another without earning a degree from any. But there the similarities with Copernicus end. Copernicus lived a fairly dull and stable life; Tycho's was wild and flamboyant from start to finish. His parents were still very much alive when his uncle Joergen, having no children of his own, kidnapped Tycho. Only after having a second son, a spare, as it were, did Tycho's parents relinquish the fight for their first born.

Joergen saw to an early and rigorous education for Tycho, sending him off at the age of 13 to the University of Copenhagen. He was to study law, a profession suited to the wealth and cultured status of his family. He left Copenhagen after 3 years with no degree. In 1562, Tycho left Denmark altogether to enroll in the University of Leipzig, again with an eye toward a degree in law, and this time including classics. Learning Latin gave him access to the *Almagest*, and being rich allowed him to buy the book for himself. It was during his stay in Leipzig, with a developing interest in the stars and planets, that the mistimed conjunction of Jupiter and Saturn clarified his role for the future of astronomy.

Tycho left Leipzig with no degree, and apparently little interest in the law. His uncle persisted, sending him next to the University of Wittenberg and then to Rostock, both in Germany. It was in Rostock where Tycho found himself in disagreement with a fellow Danish student, allegedly over the issue of who was the better mathematician. They elected to settle the dispute with swords rather than by calculation, but the duel ended when the major portion of Tycho's nose was sliced off. He survived, but the disfigurement was overwhelming, to the point that he put his metallurgical skills (an avocation) to use to manufacture a prosthetic nose. The legends had the nose made of gold or silver, or some combination of the two, but a recent exhumation of the body and analysis of the residue around the nasal cavity indicated an alloy of copper, perhaps bronze. Portraits of Tycho clearly show the metal nose in place, held fast by some compound of wax.

Tycho left Rostock with neither a nose nor a university degree. This may be one more similarity to Copernicus, both young men were pushed by uncles toward practical and lucrative professions, but were distracted by their disposition to an interest in astronomy. Tycho's obsession was observation and instrumentation, and he began building his own tools for measuring heaven. He started with a common hand-held compass, normally used for drawing sections of circles. Holding the hinged end up to his eye and sighting down each leg of the compass toward a different star, the drafting tool adapts into a handy device for determining the angular separation between objects in the sky. He started with a compass, and ended up with a small town supporting the most sophisticated astronomical equipment.

In 1575, the Danish King Frederick II granted Tycho the lordship of a small island in the Baltic Sea on which to build an observatory. Tycho called it Uraniborg, the castle of Urania, goddess of the heavens. The campus of the observatory included four observing rooms, eight bedrooms, a printing press, and a prison. It is rumored that the lord Tycho was a taskmaster and not altogether kind to his help. He would rule over Uraniborg until 1597, when he transported his astronomical tools and interests to Prague, at that time the capital of the Holy Roman Empire. Tycho Brahe died in 1601.

The flagship tools of the Uraniborg observatory were the large sextant and quadrant. A sextant is like an oversize compass with the hinged end directed up to the sky. It's attached to the floor but mounted on a ball joint to allow rotation in any direction. One leg can pivot such that with a different celestial object sighted along each of the legs, the angular separation between the two is indicated on a calibrated arc. The arc extends to 60°, a sixth of a circle, hence the term sextant. With two people viewing, one along each of the legs, the device was capable of precision to one minute of separation, a sixtieth of a degree. This was a ten-fold improvement over measurements of a generation before.

More famous than the sextant is Tycho's so-called mural quadrant. It operates on the same principle as a sextant but it is big enough to sweep through a quarter of a circle. This one had a six foot (1.8 m) radius and was fixed in place on a north-south wall of the observatory, directed to a small window at the top. A mural on the wall shows astronomers at work recording the moment when a star crosses the meridian, the great circle drawn from one celestial pole to the other and through the point directly overhead, the zenith. Crossing the meridian marks that moment when a celestial object stops rising and begins to set, the high point in its diurnal trip across the sky. The timing of this event is informative to the astronomer. The mural shows two clocks, but they have only hour hands. Tycho was apparently mistrustful of mechanical clocks, relying instead on the clockwork movement of the stars to keep time. There's more to see in the mural, details of decoration in the observatory, more tools of observation, portraits of notables, Tycho's dog. Perhaps most interesting is the celestial globe set on a shelf; it's a globe that allows for a rotating Earth.

The greater precision of Tychonic tools delivered more information from astronomical observations. A new star appeared in 1572, challenging the Aristotelian dictum that the heavens are eternal and immune to change. By carefully trying to measure the parallax, Tycho determined that this *stella nova* was indeed further away than the Moon, genuinely celestial. The challenge to Aristotelian cosmology was real. And then there was the comet of 1577 that Aristotelians would put somewhere below the Moon, in the Earth's upper atmosphere. But again, precise parallax measurements indicated that this comet, and presumably all comets, were well beyond the Moon. Not only did a comet appear and disappear, again contrary to the claim that there is no change in the heavens, but the trajectory seemed to cross the heavenly spheres of the planets. This shouldn't be possible if the spheres are made of solid crystalline quintessence.

Uraniborg is not far from Copernicus' observatory in Frombork, and similarly situated at the edge of a cold and foggy sea. The hazy circumstances did not seem to inhibit Tycho in his study of celestial objects and events. It is also not far from the site of Elsinore castle of Shakespeare's *Hamlet*, and there is some speculation that the play includes references to Tycho. *Hamlet* was first performed in 1601, the year of Tycho's death. There are numerous descriptions of astronomical phenomena, including Hamlet's famous lament at having lost his good spirit, missing even the wonder of "this brave o'erhanging firmament, this majestical roof fretted with golden fire." More specifically Tychonic is the account of "yond same star that's westward from the pole," possibly a reference to the *stella nova* that so caught Tycho's attention. A famous portrait of Tycho shows him surrounded by coats of arms of his ancestors, among them are the families of Rosencrantz and Guildenstern. The deaths of these characters in the play could have been Shakespeare's way of denouncing Tycho's model of the universe—described below—and celebrating the end of geocentric cosmology in general. Maybe. There is even some speculation that the character of Claudius, the murderous uncle, is meant to be Tycho Brahe himself. It gets complicated, but Tycho may have had an affair with Queen Sophie, Frederick's wife, and been the father of the succeeding king, Christian IV, who, to prevent the knowledge of his being illegitimate, both in birth and to the throne, had Tycho killed by mercury poisoning. Maybe. But note that Ptolemy's first name was Claudius.

Whether or not he was inspiration for Shakespearean drama, Tycho was clear in his admiration for Copernicus, referring to him as "a second Ptolemy." But he did not endorse the new world system with the Sun at the center and the Earth in motion, at least not completely. All of the terrestrial evidence, he thought, indicated that the Earth does not rotate. Tycho had nothing new to offer in the way of on-the-ground testing for rotation. He was an astronomer, after all, not a physicist. He casually mentioned the phenomena of projectiles that do not fly westward and the air and clouds not left behind as the ground, allegedly, rushes to the east. As an astronomer, Tycho could contribute to the testing of the Copernican annual revolution of the Earth around the Sun. The stellar parallax resulting from the Earth's movement from one side of the Sun to the other over the course of 6 months might now be detectable with the enhanced precision of instruments at Uraniborg.

The basic idea of parallax is easy. An object's position shifts in the field of view as the viewing location changes. You can do it by blinking from one eye to another. But measuring parallax in the sixteenth century, stellar parallax in particular, was much trickier than blinking your eyes. All of the stars, it was thought at the time, are the same distance away from the Earth. They all occupy the one celestial sphere that Ptolemy had estimated to be about 20,000 R_E away. It's not just the great distance that's the challenge, it's the uniformity. Modern astronomers measure stellar parallax by recording the shift of a nearby star against the unshifted background of very distant stars. That's what you're doing with your blinking eyes, too. The nearby object, your thumb, appears to move with respect to a more distant and stable background such as the wall of the room. There is no detecting motion, real or apparent, without some visible reference that does not move. But if all the stars are the same distance away, they will all parallax-shift by the same amount, with no visible effect. The change of position of a star, that is, the change of its angular position, must be with respect to something that does not change, or at least something that changes less, and by a known amount. Measuring parallax requires specifying the reference.

Tycho chose the pole star, what we call the north star, as his subject for a parallax measurement. In a letter of 1598, he explained to his student Johannes Kepler that observing the star near the celestial north pole would limit the atmospheric distortions. He would measure the angle between the pole star and the point in the sky around which all the stars appear to rotate, the celestial pole. From the perspective of the Copernican model, this is the angle between the star and the axis of the Earth's rotation, and it should change slightly as the Earth moves in its orbit around the Sun. That's the parallax. At the moment the pole star crossed the meridian, Tycho measured the angle between it and the zenith. The great mural quadrant was ideal for this task. Then, it's just a little straightforward trigonometry to get from this measurement to the angle between the star and rotational axis. Measuring this at a 6-month interval should reveal the parallax, if there is any, and if it is large enough to register within the tolerance of the quadrant.

It didn't. Tycho, despite honest efforts, found no evidence of stellar parallax. The standard excuse by Copernicans had been vague, simply that the stars are just too far away. Tycho asked for precision in just how far away they would have to be for the annual parallax to fall below the threshold of his measurements. It's an easy and uncontroversial calculation. The stars must be at least seven hundred times farther out than Saturn, the most distant planet. This is not just a minor adjustment to cosmology; it increases the size of the universe by nearly three orders of magnitude. But the bigger problem was all that empty space between the planets and the stars. It is especially troubling since, with the Aristotelian understanding of elements, there can be no vacuum in the universe, no genuinely empty space.

With the sharp focus of Tycho's astronomical data, the Copernican model could not hide from its empirical challenges. It's geometric elegance was attractive to Tycho, but at the prohibitive cost of making the Earth move. As he put it,

> Copernicus nowhere offends the principles of mathematics, but he throws the Earth, a lazy, sluggish body unfit for motion, into

a motion faster than the aetherial torches (the stars) and a triple motion at that.

It was satisfying, then, when he struck upon a compromise that allowed the best of both world systems.

In 1588, he published a hybrid model of the universe, a Tychonic system. It retained the essential Aristotelian requirement of purely circular orbits but it eliminated the large, looping epicycles of Ptolemy. The Earth, that lazy, sluggish body, is at rest at the center of the universe. The stars in their celestial sphere orbit the Earth once a day. The Sun also orbits the Earth, once a day, as does the Moon. So far this is basic Aristotle. The difference is that the planets orbit the Sun. As we now think of the Earth revolving around the Sun and carrying the orbiting Moon along with it, Tycho had the Sun revolving around the Earth and carrying the five planets along with it.

If you draw the orbital circles in this arrangement, it's obvious that the orbit of Mars crosses that of the Sun. This doesn't mean the two bodies will collide, but it does mean they cannot be held in solid crystalline spheres. This, and the realization that comets are celestial, made it all but impossible to claim that the spheres were physically real.

Again with an eye on the pattern of circles and revolving planets, the Tychonic model may look like a contrived and awkward mess, a desperate attempt to hold the Earth at the center of the universe and keep it from moving. But this depends on your point of view, literally. The Tychonic and Copernican models are exactly the same system, simply drawn from two different reference points, two different points of view. One chooses the Earth as its reference of being stationary; the other chooses the Sun. It is really just a choice.

To show that this is the case, let's clarify the ideas of relative motion and relative reference frame. Think of two objects in otherwise empty space; we will eventually make one of them the Sun, the other the Earth. Don't privilege either of these objects with any religious or metaphysical distinction that would require a special place or dignified status in the universe. A basic, coldhearted description of position and motion tells us that either object can be used as the starting point, the origin of the reference frame, and either one can be used as the stationary point to measure any relative motion of the other. We should be getting used to this relativity of place and motion, since navigational maps in automobiles usually show the landscape moving past, even pivoting around, as the car stays stationary in the center of the display. What's really happening, is the car moving or is the ground? The fundamental idea of relativity is that there is no real, physical difference; there are just two, equally true, descriptions of the same events.

Now put these basic observational facts to work with the Sun and Earth and describe what happens from the two different perspectives. If you choose the Earth as the origin, then the Sun must be in daily orbit, a little slower than the orbit of the celestial sphere of the stars. If you choose the Sun as the origin, then the Earth is rotating and it has a slow annual revolution that accounts for the changing backdrop of stars and the Sun's apparent position along the ecliptic. It's the same situation, alternatively described. Now just add in the planets in orbit around the Sun.

It's not that we can't detect the difference between the two models, the Tychonic and the Copernican; it's that there is no difference. There is just one world system differently described by Copernicus and Tycho.

These are the same physical systems, at least as far as we can directly observe. But if we ask about the unseen causes of motion, what holds the planets on orbit, for example, or what keeps them going, it may be that the required dynamics are different between the two systems. The solid crystalline spheres held the Ptolemaic and Aristotelian systems together in a way that there was no need to worry about forces or tethers on the planets, but the heavenly spheres seem not to exist. What holds the Earth in orbit around the Sun? This may have a satisfactory answer, consistent with the contemporary physics. What holds the Sun in orbit around the Earth? This might be more difficult to explain.

The most efficient summary of the different world systems and the state of the debate in the late sixteenth century takes advantage of our modern concepts of kinematics and dynamics. Kinematics, recall, is the description of motion; dynamics is about the causes. The geocentric system in which the Earth does not rotate and the heliocentric system in which it does are kinematically equivalent. They differ only in how one chooses the point of reference. The significant word there is "chooses." The kinematic equivalence includes both of the geocentric models, the hybrid Tychonic and the original Ptolemaic, since the latter can be turned into the former by simply adjusting the radii of all of the planetary deferents to equal the Sun's, and then positioning the Sun at the center of all the epicycles. So the kinematic equivalence is universal.

The dynamic distinction of the systems had yet to be resolved, at least in the sixteenth century. It wasn't really an issue. Celestial motion had always been regarded as natural and eternal and in that sense uncaused. Astronomy was a descriptive science, not explanatory. Physics, on the other hand, the science of terrestrial motion, did study explanations and causes. Violent motion requires a cause, and this raised a potential distinction between the two world systems. It's terrestrial phenomena, dropping stones and the ever-present air, that seemed to tell against the possibility of rotation.

Kinematics is about what is directly observable, where things are and how they move. Dynamics is about what cannot be observed. If we stick with the basic observations, as is appropriate in the context of the late sixteenth century, the early decades of the Copernican Revolution, the situation was a variation of Buridan's ass, a relativistic version. Given the kinematic equivalence of the two world systems, geocentric and heliocentric, and the competing scientific schools of thought, it's as if there is just one bale of hay but two asses staring at it. One ass insists the hay is to the right, while the other swears it's to the left. It seems they are both correct.

The next challenge is to see what happens when a robust dynamics is applied to both heaven and Earth. Will it be possible to find laws of interaction that maintain a full relativity of rotation so that a rotating Earth and an orbiting Sun are physically equivalent? This key question persisted for centuries. Here are two reputable spokesmen on issues relativistic, Albert Einstein and coauthor Leopold Infeld, raising the issue in the twentieth century:

> Can we formulate physical laws so that they are valid for all [coordinate systems], not only those moving uniformly, but also

moving quite arbitrarily, relative to one another? If this can be done, our difficulties will be over. We shall then be able to apply the laws of nature to any [coordinate system]. The struggle, so violent in the early days of science, between the views of Ptolemy and Copernicus would then be quite meaningless. Either [coordinate system] could be used with equal justification. The two sentences, "the sun is at rest and the earth moves," or "the sun moves and the earth is at rest," would simply mean two different conventions concerning two different [coordinate systems]. … Unexpected adventures still await us.

CHAPTER 7
On Skepticism

[They] seem to me to be making the mistake of judging on the basis of their own experience instead of taking into account the peculiar nature of the universe.
—Ptolemy

… idle babblers, ignorant of mathematics, may claim a right to pronounce a judgment on my work. … I consider their judgment rash and utterly despise it.
—Nicolaus Copernicus

As the history of cosmology shifts from one perspective to another, from the ancient to the modern, this is an appropriate place to reflect on the essential influence of perspective in doing science, and in doing history, come to that. Before we go to trial, Galileo's, it's worth reviewing the rules of evidence.

We can learn about ourselves by studying the past. But we also tend to describe the past in terms of how we think of ourselves. This is as true for the history of science as for cultural or political history. What we see and what we find depends in part on what we're looking for and how we look. Realizing this is the key to finding out what really happened; know your own perspective.

This is a general condition of evidence, in history and in science. The data that are used to form and test theories are unavoidably influenced by the background and context in which they are collected and used. Sometimes the influence is explicit and easy to spot; usually it is implicit and disguised.

A revolution was starting at the end of the sixteenth century, from the old world system to the new, but it was going slowly. It would take a century, from Copernicus to Newton, to reform the scientific consensus on the organization of the universe and the rotation of the Earth. Conceptual inertia is an inevitable impediment to change, since evidence, even new evidence, is at first interpreted and accredited under the influence of old theories and ideas. A stable theoretical system is a requirement for doing science, for quality control and for making sense of the evidence. In this way it is in the nature of science to impede its own revolutions.

From our perspective, the holdup on the Copernican revolution and the realization that the Earth rotates is often described as the interference by cultural and ideological forces in the process of science. Nonscientists were denying and trying to silence the work of scientists. Books were banned and individuals were punished for their support of the new world system. We point out the high-profile cases like Galileo, literally brought to his knees for his continued defense of the idea that the Earth moves. Galileo was justifiably afraid during his trial in 1633, knowing that 33 years earlier the Inquisition had seen to the public burning of Giordano Bruno for numerous crimes of heresy, probably including his outspoken advocacy of the idea that the Earth revolves around the Sun and rotates on its axis.

There is no denying the clash between Catholicism and the Copernicans, but it is misleading to blame only this for the slow and difficult progress in the acceptance of the new world system. From here we see similarities to the present, when important matters of science, things like the explanation of global warming and even the fundamentals of evolution, are denied despite clear evidence and consensus among experts. Idle babblers, ignorant of mathematics and theory, claim the right to pronounce judgment on everything from climate change, to the safety and efficacy of vaccinations, to the value of genetically modified foods. These are important issues, and in the sixteenth and seventeenth centuries, so was knowing the location and stability of the Earth. It's tempting to group all these cases together as examples of the denial of science, motivated by some cultural need to promote or retain an ideology. It is tempting and it will be informative to look, but we have to be mindful of the significant differences between our case and the Copernican revolution.

The context of scientific struggles in the sixteenth and seventeenth centuries was notably different from what we now describe as science denial or even a war on science. In 1600, when Giordano Bruno met his fiery end in the Campo de'Fiori in Rome, and when Galileo's challenge to scripture was being drafted, the scientific community was not of one mind on issues of cosmology. We can't accuse the Catholic church of suppressing the authority of science on the issues of the rotation of the Earth, because at the time there really was no single scientific authority on that issue. It was a time of genuine scientific disagreement.

Consider the case of Giordano Bruno, who draws our attention in part because of the drama of his demise. It's irresistible to consider him a martyr for the cause of science, and being burned alive makes him a good candidate for martyrdom. But it's a stretch to say he was killed for his scientific views and that it was done to prevent the spread the new world system. He was indeed outspoken in his endorsement of the Copernican model from as early as 1583. In a series of public lectures given at Oxford, he was clear in saying that the Earth really rotates. He offered no new evidence that the Earth moves, but his reasoning went beyond the humble notion that this was not the center of the universe. Not only is the Earth not at the center of things, neither is the Sun. In fact, there is no center at all, and that's because the universe is infinite. Copernicus described a universe of enormous size but still a sphere of determinate, finite diameter. Bruno, arguing that the creative capacity of God was unlimited, concluded that the size of the creation must match the ability of the creator. He went on; as the Earth orbits the Sun, and the Sun is a star similar to the untold number of other stars that brighten the night sky, there must be other planets orbiting those other stars, other worlds such as our own.

Giordano Bruno was never shy to challenge authority and doctrine, and he was rarely tactful in his manner. Not surprisingly, this aspect of his personality did not serve him well as an ordained Dominican priest. In 1572, while at the convent of San Domenico de Maggiore in Naples, he openly questioned the divinity of Christ and was thereby excommunicated and threatened with a trial for heresy. He escaped to Rome, and thus began a lifetime of itinerate offense, touring the intellectual centers of Europe, staying just the short time it took for his combative and confrontational style of critique to get him run out of town. He went north to Geneva, where he embraced Calvinism, but a published attack on a spokesman of the church got him arrested and excommunicated. He escaped to Paris, and then in 1583 to Oxford where he presented his ideas on cosmology. The Oxford scholars were not receptive to the radical claims of many worlds in an infinite universe, let alone a moving Earth and Bruno returned to Paris. Confrontation with individuals of the intellectual elite, and a published assault on Aristotelian science, forced a leave from France. In Germany, he became a Lutheran. A series of short-term teaching jobs in a variety of German university towns brought him to Helmstedt, where his disregard for critical decorum got him excommunicated from the Lutheran Church. Giordano Bruno is thus distinguished by being excommunicated from all three of the major churches of Renaissance Christianity, a heretic's hat trick.

From Germany he returned to Italy. It was a fatal mistake. Venice seemed a safe place to bring his polemic back home, as it was the most progressive of Italian states. But it wasn't long before dispute with a benefactor had Bruno reported to the Venetian Inquisition and extradited to Rome where the wandering antagonist was finally brought to face the charges of heresy. The specific counts of indictment have been lost, allowing historians ongoing disagreement on whether it was a case of religion versus science or a more intramural affair of the Roman Catholic church disciplining one of its own over transgressions of faith. In light of Giordano Bruno's lifetime of tactless contempt for authority, both religious and scientific, and noting that he was initially charged with heresy in 1576, 7 years before his public endorsement of a moving-Earth cosmology, it would be simplistic, and probably an imposition of our own situation, to call this a case of a cultural attack on science.

Galileo's run-in with the church offers more evidence to the historian, but there remains significant disagreement over the motives and issues in play. The spectacle of a trial draws our attention in a way that may be a distraction from the broader understanding of the time. It's just one case, and though it is easy and entertaining to make it the exemplar of a fraught relationship between science and religion—and I have been as guilty as anyone in this—it must be considered in context. There is no denying that Galileo's endorsement and teaching the idea that the Earth moves was the basis of the charge of heresy. It's in the indictment, the transcript of the trial, and the verdict—the condemnation by the Inquisitors. An influential organization of nonscientists was systematically opposing an important scientific result. It's fair to claim that similarity to the present. But it's equally important to point out some significant differences.

Consider the case of climate change and the frustration among scientists with the resistance by some powerful political and economic forces. Scientific studies of scientific studies report an overwhelming agreement among scientists that global warming is real and is caused by human production of greenhouse gases. The consensus is the clearest evidence to convince the public and the politicians of the accuracy of the claim. Don't talk about the evidence directly; talk instead about the authoritative verdict. Ninety-seven percent of climate scientists agree, and to oppose or doubt their findings is an act of science denial. At the start of the seventeenth century, there was nothing like 97% endorsement of the new world system, and this makes our situation today much different from what was happening during the Copernican revolution.

In the late sixteenth and early seventeenth centuries, when Bruno was burned at the stake and Galileo was confined for life, there was no single scientific consensus that the Earth moves. Cosmology was a science divided. The heliocentric and geocentric models of the universe each claimed a school of followers, but tradition, scientific tradition, still favored the old world system. Opposition to either model was from the scientific base of the other, using science rather than denying science. At the time, the evidence was not conclusive. There were legitimately two world systems to consider, significantly unlike the unified understanding of climate change or the one fundamental principle of natural selection presented by science in our time. There were two schools of thought, separated consensus. Each of the world systems was connected in its own coherent network of ideas of nature and would be able to use its own theories to interpret data. What you understood about natural motion or the relation between heaven and Earth influenced what the basic observations indicated about the rotation of the Earth.

This is how science works, and it's part of what makes science scientific. Evidence is more than haphazard observation; it's carefully gathered and interpreted observation, using the best theoretical understanding to certify and make sense of what is observed. This dependence on background knowledge gives the testing of theory by evidence a whiff of circularity and potentially makes the process self-affirming. It's unavoidable. Israel Scheffler, a philosopher of science, put it with blunt clarity. "Observation contaminated by thought yields circular tests; observation uncontaminated by thought yields no tests at all." But it is not hopeless. We need to figure out how, in the case of the rotation of the Earth and generally in science, evidence influenced by theory can be used as an objective guide to know in fact whether the Earth rotates.

Consensus on a scientific conclusion is today cited as a virtue, a good reason to trust the results and believe they are true. But at the beginning of the Copernican revolution, consensus was an impediment to change, since the lingering authority of the Aristotelian science, the canon of the scholastics, confounded the revolution. This highlights the tricky balance between authority and novelty in science. Consensus is the enforcement of authority; it is the enemy of change. Science has this built-in inertia, a self-contained standard of legitimate ideas that was the primary impediment to change in the sixteenth and seventeenth centuries. The conflict was not so much a cultural rejection of science as an intramural dispute between two scientific descriptions of the universe.

Modern science is proud to point out the overwhelming consensus on global warming. It's a hard-won agreement, a trial of rigorous testing and dedicated respect for evidence, with a jury of tough-minded skeptics. But there are components of the scientific process to consider before awarding all debates to the prevailing opinion of scientists. Peer review of publications and the allocation of research funds is the bedrock of quality control in science. Crackpot ideas or sloppy procedures won't make it into the scientific marketplace of ideas, and this is as it should be. Otherwise, not much good science would get done amidst the overwhelming rabble of anything-goes publication. A lot of time and money would be wasted. But clearly, the important role of peer review promotes agreement in science and makes consensus easier. Conformity is rewarded, while novelty is turned away.

The culture of conformity is built into the structure of science education. The Socratic method of leading questions to get students to arrive at answers on their own just doesn't work in the natural sciences like physics and astronomy. There are facts to be taught, and the efficient means is by lecture and textbook. Thomas Kuhn, the historian of science who brought the ideas of a paradigm and paradigm shift into the contemporary vernacular, pointed out the unique role of textbooks in the education of a scientist. "The single most striking feature of this education is that, to an extent totally unknown in other creative fields, it is conducted entirely through textbooks." This is how peers are produced, by everyone learning the fundamentals through the clear, confident presentation in textbooks. And this is as it should be, again to provide the stable foundations on which to build and against which to evaluate methods and results. The education of a scientist is the preliminary initiation into the profession, and like peer review it works to encourage conformity and promote consensus.

Scientific consensus, in other words, has an enigmatic role in the process. It can both underwrite truth and perpetuate falsehood. Consensus alone cannot make the case that the agreed-upon ideas are accurate. It is, however, a sign of the robust network of background knowledge that is necessary for making a credible case, necessary for the careful collection of data and its meaningful use as evidence. This kind of stable base of theoretical support is generally missing when nonscientists challenge and deny the results of research. These are hit-and-run skeptics, idle babblers, with no understanding of the scientific context, and just as importantly, no alternative conceptual context of their own. Responsible science requires a coherent theoretical structure in support of each conclusion; responsible skepticism requires nothing less.

Tom Nichols describes this as "the death of expertise," lamenting the electronically enabled crackpot to blog or comment on important matters about which they know almost nothing. The commentary usually amounts to objection with no real justification. It ignores the fact that in science, unlike in a courtroom trial, both sides of any issue must carry a burden of proof. A clear symptom of irresponsible skepticism is cherry-picking the data. A cold snap does not disprove a general warming of the climate. A case of autism diagnosed after vaccinations does not prove a casual connection. Such shallow interpretation of data reveals a lack of systematic interpretation, whether it's using your own system or some other. It's idle babbling.

Ptolemy and Copernicus would agree on this assessment of outside commentary on science. For Ptolemy, one's personal experience is no credible challenge to the considered conclusion of scientists who know "the peculiar nature of the universe." They are guided by a robust theoretical understanding of nature. Copernicus is more focused but ready to dismiss the objections of anyone "ignorant of mathematics." In either case, it's peer review that counts, not, using an anachronistic analogy, anyone-with-a-blog review. Their sentiment can be extended to the context of the classroom. In the current controversy over teaching evolution in public schools, one proposed solution is to teach both evolution and some version of intelligent design and let the kids decide for themselves. This sounds evenhanded, but school children generally lack the system of background knowledge to make a responsible decision about any scientific theory. Their nonexpert opinion will only be an illusion of informed consent, whatever their conclusion.

External and independent review of ideas is of course a good idea, and this would include the view from outside the textbook-initiated peer group. Think outside the box. Indeed, but there can be no thinking outside of at least some box. The box is a metaphor for the guidelines of good reason and the conceptual and theoretical basis for meaningful, credible evidence. Nonexpert criticism of scientific results are most often not just outside *the* box, but outside any box whatsoever. Again, idle babble.

Ptolemy's concern about judgment based on one's own experience is somewhat ironic, particularly for someone who believes the Earth stands still, as we all experience. But it gets to the crux of the issue of the role of evidence in science, and in day-to-day life. We are being asked, certainly in matters of astronomy and physics, but also in matters of health, nutrition, climate, and pretty much all that matters, to believe things that we haven't observed for ourselves. More of a challenge, we are being asked to believe things we can't observe for ourselves, and in fact no one can observe directly. We are even expected to believe claims about the natural world that are contrary to what we, and everyone else, observes. In the sixteenth and seventeenth centuries, this included the uncontroversial fact that the Earth is round and the more contentious claim that it rotates. This is why we need clear rules of evidence, rules that are prescribed and enforced by the people who understand the peculiar nature of the universe, that is, scientific peers.

The most important rule is to acknowledge that all evidence is from one perspective or another and bears that indelible influence. The perspective is both physical and conceptual. There is the role of the physical reference frame, the choice of what to use as the stationary framework to detect and describe position and motion. There is no observation of motion other than by reference to something else, something held still. And there is the role of the conceptual reference frame, the theoretical box, the one in which we think, directing the selection, certification, and significance of evidence. Israel Scheffler, again, "observation uncontaminated by thought yields no tests at all."

How is it that observation contaminated by thought, that is, evidence influenced by perspective, can be used to test new ideas without simply imposing the old ideas? The key is to make the perspective explicit and to admit that it's not the only possible perspective. On the physical influence of a reference frame, keep in mind that any observation of motion must include a reference, explicit or implicit, to some stationary object of reference. That's a chosen perspective. On the conceptual influence of accepted physical theory, the textbooks and peers, keep in mind that the interpretations of the data are dependent on a prior understanding of the important physics of natural motion and basic metaphysics like the allowable shape of celestial trajectories.

Dogma is the condition of holding fast to the entrenched conceptual scheme that is hard at work in interpreting evidence, refusing to allow that the network of ideas could be false, in part or entirely. Denial is more casual, using only a shifty core of interpretive principles, allowing piecemeal or convenient doubt. In either case, the disservice to the pursuit of knowing what's really going on in nature is the failure to carry the full burden of proof for claims both positive and negative.

Neither dogma nor denial describes the situation at the beginning of the seventeenth century and disagreements about the rotation of the Earth. It's telling that Galileo described the Ptolemaic and Copernican cosmological models as world systems. It properly indicates that both models of the universe were packaged in comprehensive and coherent networks of theory about, quoting Ptolemy yet again, "the peculiar nature of the universe." Advocates of each model accepted the obligation to test and the possibility of their own conclusions being wrong. Aristotle was a clear role model in the scientific tradition of keeping in mind the possibility that he was wrong, and seriously considering an opposing hypothesis. It is also telling that Galileo presented the science in the form of a dialogue, and in the vernacular Italian. This is the essential balance of authority and challenge, inviting anyone, peers or not, to consider the two sides.

This gives us the opportunity to answer the question about theory-influenced evidence resolving disagreements about theory, how observation contaminated by thought can test new and controversial ideas, in the specific case of the rotation of the Earth.

A coherent network of beliefs is a necessary condition of an accurate world system. The challenge, and the indication that the descriptive claims in the system are true is an ongoing expansion of the network—more data and more phenomena explained—while maintaining consistency. This is what will break the impasse in the seventeenth century and award the revolution to the Copernicans.

At the time of Copernicus and Tycho, astronomy was pure kinematics. There was little interest in the causes of heavenly motion, no celestial dynamics. The modern perspective, initiated by Galileo and Kepler, began to merge terrestrial physics with celestial astronomy in a way that demanded dynamics. Physics asked about the causes of motion, and the loss of the stabilizing heavenly spheres, the quintessential globes that held planets and stars in place and carried them around in revolutions, necessitated explanations of what held the celestial objects up and kept them going. Kepler would make things worse before he made them better, by suggesting that the shape of planetary orbits is not circular or spherical but elliptical. The eternal repetition of following a circle was lost, and the motion could no longer be seen as natural; this violence in the heavens required a force. One of the chief world systems would survive the expansion into dynamics; the other would not.

Dynamics is a science of causes. It deals in things that cannot be directly observed, forces like gravity and electromagnetism. This makes any dynamical theory vulnerable to change, with new evidence or a new way of thinking. That's why, after the modern perspective of Galileo and Kepler and Newton there will be a post-modern perspective of Einstein and theories of relativity. Newtonian dynamics will have an impact on interpreting data about the rotation of the Earth. So will relativistic dynamics.

The Modern
Perspective

CHAPTER 8
The Two Chief World Systems

And yet it moves.
–Galileo

There is good reason to believe that Galileo never read beyond the first chapter, Book One, of the book by Copernicus, *On the Revolutions of the Heavenly Spheres*. He seems to have paid little attention to either the messy details of epicycles or the intricate geometric exercises that complicate the rest of the book and make it more challenging reading. He skipped the math, despite being a mathematician, and he seemingly ignored his own, now famous, advice.

> Philosophy is written in this grand book, the universe, which stands continually open to our gaze. But the book cannot be understood unless one first learns to comprehend the language and read the letters in which it is composed. It is written in the language of mathematics, and its characters are triangles, circles, and other geometric figures without which it is humanly

impossible to understand a single word of it; without these, one wanders about in a dark labyrinth.

It's not just that he didn't share the details with his readers, or burden them with the math. His own copy of Copernicus has almost no annotation or marginalia beyond the first chapter. He was careful to edit out the offensive bits as instructed by the Inquisition, for example striking through "The Explication of the Three-Fold Motion of the Earth," and penciling in the approved, "The Hypothesis of the Three-Fold Motion of the Earth and Its Explication." But beyond this compliance with Catholic censorship, there is nothing in Galileo's hand to indicate an attention to detail.

What convinced Galileo, and what he presented to win over his readers, was a simple, first-approximation version of the heliocentric model of the cosmos. The Earth revolves around the Sun and rotates on its own axis—two motions, though clearly Copernicus described three. There are no epicycles in Galileo's version, and he suggested that none are needed. "Ptolemy introduces vast epicycles, ... all of which can be done away with by one very simple motion of the earth." He seemed also to ignore the possibility of elliptical orbits, that radical revision introduced by his German contemporary Johannes Kepler. Galileo put the planets on perfectly circular orbits, centered on the Sun, and opened a dialogue on the model of the universe pretty much as we understand it today in elementary school. The Catholic church couldn't have suspected the devil was in the details, because there were no details.

The grand book of nature may well be written in the language of mathematics, and impossible to understand otherwise, but apparently Galileo's readers would be expected to believe what they read, even without troubling to do the math. Despite Copernicus' disapproval, judgment could be passed even by those ignorant of the mathematics.

Galileo didn't start out his intellectual life studying mathematics. As the eldest child in an academic family—his father Vincenzo Galilei was a musician and musical theorist—he began his education early. In 1574, when he was ten years old, he entered school at the Camaldolese monastery near Florence. He found the monastic life agreeable and considered staying to join the order, but his father had other, more practical plans for the son who would eventually be financially responsible for the family. Galileo was sent to the University of Pisa to study medicine. This was apparently less to the son's liking. His studies gave Galileo an introduction to mathematics, and in 1582, Ostilio Ricci, a visiting scholar, connected the teenager with his life's calling. Galileo dropped out of med school and became a freelance mathematician, an itinerate tutor.

Mathematics at the time was securely founded on the Greek classics. Euclid's 13 volume treatise on geometry, simply titled *Elements*, was the core. Geometry, after all, is exactly what you need to make sense of the orbits, equants, and epicycles in the Ptolemaic universe. Its value was, and still is, not just for the comprehensive analysis of lines, angles, circles, and solids, but in the immaculate organization of reasoning that links the concepts and derives new ideas from old. With its foundational structure of definitions, axioms, and proofs, Euclidean geometry is the model of precise, deductive reasoning. One thing follows from another with genuine certainty. That's why Aristotelians allowed mathematics in describing celestial phenomena, and disallowed uncertainty in any science. It's also why most of us were required to take geometry in high school, for the mental calisthenics.

Other familiar Greeks were featured in the education of a sixteenth-century mathematician. Pythagoras—whose followers, you will recall, claimed that the Earth rotates—was a pure mathematician, almost metaphysical in his regard for numbers. Archimedes, by contrast, was a patron of the applied. He put numbers to work in pumping water, leveraging heavy weights, and destroying invaders' ships. He offered less practical calculations as well, for example, figuring the number of grains of sand it would take to fill the universe. Eight vigintillion, it turns out. That's eight followed by 63 zeros. Of course this is an approximation, and that's important, allowing mathematics to deliver approximate and uncertain results. For both the applications and the uncertainty, Archimedes was a role model for Galileo.

There's money in math if you can get the numbers to solve real-world problems and meet material challenges. Galileo contributed to future war efforts with calculations of the distance a cannon ball will fly, depending on the angle it is launched. He was also involved in the fine-tuning of the recently adopted Gregorian calendar. And, by invitation of the Florentine Academy, he delivered a lecture on the dimensions of hell in Dante's *Inferno*. Just 1 year later, in 1589, he got a full-time job as a lecturer in mathematics at the University of Pisa. Three years after that, he tripled his salary by moving to the University of Padua as a professor of mathematics. He stayed in Padua for 18 years, later saying it was the happiest time of his life.

The division of disciplines at a Renaissance university had astronomy taught by mathematicians and physics taught by philosophers. Galileo, with an interest in falling stones and the trajectory of a cannon ball, worked in what we might call inter-disciplinary studies. He applied math to physics and proposed the relevance of physics in the investigation of what happens in the sky. When he left Padua in 1610 to take a job in the court of Cosimo II de'Medici, Galileo insisted his title include the word "philosopher": Mathematician and Philosopher to the Grand Duke of Tuscany. After 6 years as mathematician and philosopher to the grand Duke, Galileo retired to a home near Florence. In 1632, he published the *Dialogue Concerning the Two Chief World Systems*.

Even when you merge the disciplines of astronomy and physics, there remains an important distinction between celestial and terrestrial phenomena, and there is a difference in how they can or cannot provide evidence that the Earth rotates. Galileo was explicit in the *Dialogue*, that no experiment or natural events on the Earth itself can demonstrate rotation. Salviati, spokesman for the Copernican model, explained, "whatever motion comes to be attributed to the earth must necessarily remain imperceptible to us and as if nonexistent, so long as we look only at terrestrial objects." The only evidence that will reveal the real motion of the Earth will be from the apparent movement of the stars and planets. Look to the sky to see if the Earth is moving, as you look out the window of the train and watch the scenery go by to know that you are moving.

This is the setup for one of Galileo's demonstrations that the Earth rotates. The evidence will not be conclusive, so there will be more to come, on the way to preponderance of good reason to believe that the Earth moves. This first reason appeals to simplicity

> The true method of investigating whether any motion can be attributed to the earth, and if so what it may be, is to observe and consider whether bodies separated from the earth exhibit some appearance of motion which belongs equally to all. For a motion which is perceived only, for example, in the moon, and which does not affect Venus or Jupiter or the other stars, cannot in any way be the earth's or anything but the moon's ...
>
> ... Now there is one motion which is most general and supreme over all, and it is that by which the sun, moon, and all other planets and fixed stars—in a word, the whole universe, the earth alone excepted—appear to be moved as a unit from east to west in the space of twenty-four hours. This, in so far as first appearances are concerned, may just as logically belong to the earth alone as to the rest of the universe, since the same appearance would prevail as much in the one situation as in the other ...
>
> ... Now if precisely the same effect follows whether the earth is made to move and the rest of the universe stay still, or the earth alone remains fixed while the whole universe shares one motion, who is going to believe that nature (which by general agreement does not act by means of many things when it can do so by means of few) has chosen to make an immense number of extremely large bodies move with inconceivable velocities, to achieve what could have been done by a moderate movement of one single body around its own center?

In other words, the Copernican world system, with a rotating Earth, is simpler than the old world system with its orbiting celestial spheres of stars and planets.

This is a frequently used standard for evaluating scientific theories; when the evidence is equivocal, opt for the simplest explanation. Much less frequent is any explicit justification of the fundamental principle showing that a simpler theory is more likely to be true. Simplicity has undeniable pragmatic virtues; anyone would prefer to work with a model with fewer moving parts if it matches the data with accuracy equal to other, more complicated schemes. But what does simplicity have to do with truth? It's not enough to say it's because nature is efficient and so it just is simple. That, as the philosophers say, begs the question. Our understanding of nature has been guided by a principle of accepting simple explanations, so of course we've come to believe that nature is simple.

Aside from the challenge of finding a real connection between simplicity and truth, Galileo's rhetorical question, "now who is going to believe …," is a long way from actually observing the Earth move. It's a vulnerable inference, making the evidence very circumstantial. And don't forget that Galileo was working with an artificially simplified version of the new world system, one that leaves out the epicycles of motion of the planets and the swirling track of the center of planetary orbit. It's not really a fair comparison with the geocentric model of Aristotle and Ptolemy. You might even argue that he misleads by counting each star individually in the "immense number of extremely large bodies." The stars, according to the old world system, are fixed objects on just the single celestial sphere. It's just the one sphere that orbits the Earth.

So, it's not obvious which model is the simpler. It almost never is. When you look at the details, arguments about simplicity are generally idiosyncratic and, well, complicated. Ptolemy, recall, had anticipated this appeal to parsimony, and pointed out the unreliable caprice in judging simplicity. In his words

> Rather, we should not judge 'simplicity' in heavenly things from what appears to be simple on earth, especially when the same thing is not equally simple for all even here.

In one way, Galileo ignored the details of the new model of the solar system; in another way he magnified them. The view through his telescope showed features of celestial objects that he presented as evidence that the Earth moves. It was not in the "motion which belongs equally to all [the stars and planets]," but in surprises seen individually in planets like Jupiter and Venus, and in the Sun and the Moon.

Galileo supplemented his professor's salary as an inventor. A military compass, for example, and a water pump that revised and improved on the Archimedes screw were among his lasting and lucrative technological contributions. In 1609, he learned of an invention in the Netherlands, a combination of optical lenses aligned in a tube that produced a magnified view of things far away. He quickly built one for himself. It was by trial and error, without any real understanding of how the lenses worked or the general nature of light. His first functioning telescope magnified what he saw by three times. With experience and several more models of the instrument, he had the magnification up to 20.

The telescope had obvious practical benefits. Threatening armies or navies were now visible at greater distance, and merchant ships sailing into Venice could be seen and anticipated in a way that allowed traders to take advantage of advanced knowledge of the market. Galileo profited from the invention. His salary was significantly magnified, and he cleverly sold the patent to the Venetian Senate, never mind that he had not invented the telescope himself.

Despite its verifiable accuracy when magnifying familiar things like ships and distant buildings, there was lingering skepticism about the telescope's use in astronomy. It seemed to manufacture and distort images, showing two stars, for example, where there was really only one. Martin Horky, a German student of astronomy studying in Bologna, gave the new instrument a try when Galileo came to town. He reported

> I tested the instrument of Galileo's in a thousand ways, both on things below and on those above. Below it works wonderfully; in the heavens it deceives one, as some fixed stars are seen double.

It made some scientific sense that a device would function differently with objects on the Earth than it would with the heavens. Terrestrial laws are fundamentally different from celestial laws, according to the science of the time. Physics and astronomy were still separate departments at the university. Aristotelian principles still structured the inquiry.

For those who endorsed the celestial application of the telescope, there were exciting new discoveries, but none of them included a view of a rotating Earth. The uneven surface of the Moon and moving blotches on the Sun challenged the old-world notion of perfection in the sky. A full range of phases in Venus implied that the planet must orbit the Sun rather than the Earth, thus eliminating the possibility of the Ptolemaic model. But the Tychonic model was still possible, and that had the Earth unmoving at the center of things. The telescope revealed moons in orbit around Jupiter, showing the possibility of a moving planet holding onto orbiting moons. If Jupiter could do it, so could the Earth. All of this collected into a body of evidence against the geocentric model of the cosmos and for the heliocentric, the model in which the Earth rotates. None of it was an actual observation of the rotation of the Earth. It wasn't even direct evidence of the Earth's rotation, in the sense of being an immediate effect of that rotation.

The evidence from the telescope did not play a big role in the *Dialogue Concerning the Two Chief World Systems*, the book in which Galileo made the case that the Earth moves. Celestial data were generally pretty quiet throughout the conversation. Except for the argument about the simpler explanation of the apparent motion of the stars, the dialogue was about what happens on the Earth itself. No new evidence was presented; old evidence was reinterpreted, including experiments cited by Aristotelians that allegedly prove the Earth is stationary. They prove no such thing, according to the interpretation by Galileo. Not only is terrestrial evidence incapable of showing that the Earth moves, it can't show that the Earth *doesn't* move, either.

The first task was to debunk a long-standing and influential demonstration against the rotation of the Earth, Aristotle's argument of a projectile launched straight up or dropped straight down. Galileo fairly paraphrased it in the *Dialogue* ...

> Aristotle says then that a most certain proof of the earth's being motionless is that things projected perpendicularly upward are seen to return by the same line to the same place from which they were thrown, even though the movement is extremely high.

This, he argues, could not happen if the earth moved, since in the time during which the projectile is moving upward and then downward it is separated from the earth, and the place from which the projectile began its motion would go a long way toward the east, thanks to the revolving of the earth, and the falling projectile would strike the earth that distance away from the place in question. Thus we can accommodate here the argument of the cannon ball as well as the other argument, used by Aristotle and Ptolemy, of seeing heavy bodies falling from great heights along a straight line perpendicular to the surface of the earth.

Galileo allowed the Aristotelian spokesman to make the case even stronger by citing a law of nature, right out of the physics book, that implies that a dropped stone not only *will* fall some distance to the west on a rotating Earth, but it in fact *must* do so.

For to expect the rock to go grazing the tower if that were carried along by the earth would be requiring the rock to have two natural motions: that is, a straight one toward the center, and a circular one about the center, which is impossible

By the laws of Aristotelian physics, the physics taught in universities and enforced by peer review at the time, an object can have only one natural motion. For celestial bodies, the natural motion is circular. For terrestrial things like stones, the natural motion is a straight line, up, or down. Moving a stone by hand, for example, carrying it as you move horizontally, is an act of violent motion. On the hypothesis that the Earth is rotating, and you are rotating with it, the stone is being violently moved horizontally as you hold it at the top of the tower. The moment you let go, the cause of violent motion is lost and the stone no longer moves horizontally. You and the tower keep going east, but the stone reverts to its one natural trajectory, straight down. It must fall behind.

Note that the logical form of this argument is exactly our current textbook rendition of good scientific testing. It's set up to falsify a hypothesis. On the assumption that the Earth rotates, we predict that a dropped stone will fall to the west. The observation is that it falls straight down, and this disproves the assumption.

The data were not disputed; a dropped stone does in fact fall straight down. It grazes the side of the tower. And there was no suggestion that the deflection is simply too small to measure. That was the excuse for not seeing stellar parallax; it's there, but undetectable because the stars are so far away. But if the stone is dropped from a tower 50 meters high, approximating the height of the leaning tower of Pisa, it takes about 3 seconds of free-fall to hit the ground. In that time, a rotating Earth would carry the tower and the ground it stands on a little over a kilometer to the east. The stone, left behind, would fall that distance to the west. That's not just detectable, it's dangerous, and it clearly doesn't happen.

There are stories of Galileo dropping things off towers. The most famous account has him at the top of the campanile in Pisa, releasing two stones, one heavy and the other light, to demonstrate to the audience below that the two fall at the same speed and hit the ground simultaneously. He has a reference himself, in the *Dialogue*, to dropping two birds, one dead and other alive, from the tower. But this is almost certainly a joke, since he follows it with a plan to drop two cats, one dead and the other alive. It's unlikely he ever did any of this, including the stones of unequal weight. He got the idea that the rate of free-fall is independent of weight by observing hail stones. A hail storm is a mix of large and small stones that he assumed were formed at the same time and at the same height. If heavy things fell faster than light, the largest hail would arrive first, followed by pieces of diminishing size. That's not what happens; the storm is a mix of large and small hail stones from start to finish. Once he had the idea of equal speeds, he finessed the proof by logic, with no need of an experiment. You can read all about it in his last book, *Two New Sciences*, written while under house arrest.

Aristotle's appeal to the vertical trajectory of a projectile, rewritten by Galileo as simply dropping a stone from a tower, is a sophisticated version of the very basic observation that it doesn't feel like the Earth is rotating. I know what spinning feels like, and this isn't it. Galileo's challenge is to accept the sensation but interpret it differently, explaining why this is exactly what to expect on a spinning Earth, stones that fall straight down. He needed to show that the stone dropped from a moving tower will not fall behind. It's a huge task, since it will defy one of the laws of nature as written in the textbook. And then it's just a first step, since this won't prove the Earth rotates. It will only show that a perennial go-to case against rotation proves nothing. It will, at least, take some pressure off the Copernican model.

In Galileo's presentation of what historians of science sometimes call the tower argument, he allowed the Aristotelian spokesman to further support his claim that a stone dropped from a moving tower would fall behind by citing an actual demonstration with similar results. It's a setup by Galileo, to make it easy to expose the mistakes, both in method and results. There is a well know experiment done on ships, says Simplicio, the Aristotelian, in which a stone dropped from the top of the mast

> falls to the foot of the mast when the ship is standing still, but falls as far from that same point when the ship is sailing as the ship is perceived to have advanced during the time of the fall, this being several yards when the ship's course is rapid.

But, of course, it doesn't. Simplicio has never done the experiment himself, and it is revealed that he is only trusting the authority of others to report the results. This is exactly what's wrong with the methods of Aristotelian scientists, Galileo points out, the scholasticism of doing science in the library. Their convictions are based on reading the words of scholars rather than observing the world for themselves. "Our discourse must relate to the sensible world and not to one on paper." There is a similar concern today. We are quick to Google, reluctant to think or look for ourselves.

In fact, we've all done a version of the experiment that Simplicio is describing, in an airplane or on a train, or, as Galileo points out, on a boat, not at the top of the mast but comfortably and safely inside a cabin, out of the wind. As long as the sea is smooth, anything dropped will fall straight down whether the boat is moving forward or tied up and stationary. With the windows closed, just attending to what happens inside the cabin, there is no way to tell if the boat is moving or not. Every experiment will have exactly the same result on a (evenly) moving ship as on a stationary one. We now refer to this fact as the principle of relativity. In modern terms, no experiment can distinguish one uniformly moving reference frame from another. Motion is relative.

There is an important difference between Galileo's conclusion from the ship experiment and our more modern, actually postmodern, principle of relativity. Galileo was pointing out that uniform motion is undetectable, but he still allowed that the ship really is either moving or not—absolute motion. The updated version deepens the relativity. Uniform motion cannot be detected because any reference frame, the cabin on the boat, for example, can be considered at rest—there is no absolute motion, only relative. We have work to do on this deeper version, and we'll do it from the postmodern perspective. For Galileo, his principle of relativity is at the heart of his initial warning that no terrestrial evidence would show that the Earth moves.

If there is any doubt about the outcome of the experiment on the boat, or lingering worry about the falling object having two natural motions, in violation of the law, Galileo made the case by logic alone. "Without experiment, I am sure that the effect will happen that way." It's a thought-experiment, with idealized conditions, using physics to draw an astronomical conclusion. This was revolutionary. It's also pretty easy. Imagine a perfectly round ball on a perfectly flat, horizontal surface. If the surface was tipped up, a rolling ball would slow down, stop, and reverse course. If the surface was tipped down, a rolling ball would speed up without end. But this surface is horizontal, so once a ball gets rolling by whatever violent means and is then left alone, it will neither slow down nor speed up; it will just continue to roll at the same speed without any external push. This is inertia, although Galileo didn't use that word. These simple logical steps show that the rolling ball, characteristic of any object on the Earth, can have an unforced horizontal motion. That is, it can have a horizontal natural motion in addition to its vertical natural motion. The law restricting natural motions to just one is wrong, and really, we knew this all along. The thought experiment just made it clear.

Galileo was not challenging the distinction between natural and violent motion. And, in fact, this distinction is still with us. Physics now puts it in terms of a so-called free particle and one with external forces acting. The free particle will follow its natural, inertial course. Galileo's dispute is only with the Aristotelian idea that each object has just one distinct natural motion, and that it is different for celestial and terrestrial bodies. The horizontal motion of a terrestrial body is really circular, as it follows the curvature of the spherical Earth. Galileo brought this celestial natural motion down to Earth, and merged astronomy with physics. It's fair to call this the beginning of astrophysics.

Putting all the pieces together, and getting back up to the top of the tower, the dropped stone, like the rolling ball, will continue its horizontal motion even after it's released. On a moving Earth, the stone will follow the ground and fall straight to the ground, landing at the base of the tower. Seeing this happen is not proof that the Earth is stationary. As Galileo warned, evidence of events on the Earth itself would not prove, one way or the other, if the Earth rotates.

The *Dialogue Concerning the Two Chief World Systems* does offer one sustained argument to show that the Earth rotates. The original title of Galileo's controversial book was *On the Ebb and Flow of the Sea*. Catholic censorship insisted on the change, apparently to highlight the hypothetical nature of the new world system by explicitly labeling it as a dialogue. Nonetheless, Galileo persisted with a lengthy account of the ocean tides, the ebb and flow of the seas, presenting this as the best evidence that the Earth moves. The Fourth Day of the *Dialogue*, the last day of discussion, is dedicated to an explanation of the tides, an explanation that Galileo claimed requires the Earth to both rotate and revolve around the Sun.

This is surprising, and a bit awkward, since Galileo had explicitly insisted that the evidence for the Earth's rotation cannot be found in phenomena or experiments done on the Earth itself. The tides are a terrestrial phenomenon. Using the tides as evidence of rotation seems to ignore his own warning and to violate his own principle of relativity that was so carefully proven earlier in the *Dialogue*.

The tides were not a big concern to Greek scientists, perhaps because the effect is generally small in the Mediterranean sea. There are some notable local exceptions, where the orientation and shape of a bay or channel focus and enhance the highs and lows. The narrows between the island of Euboea and mainland Greece is one case. The tidal bore is visible as it flows and eddies near the town of Chalcis. That's where Aristotle retired and died, and there's a rumor that he killed himself by leaping into the swirling water, frustrated by inability to explain what was going on. It's just a rumor. Another location for noticeable tides is Venice, where Galileo spent a lot of time when he lived nearby in Padua.

The history of explanations of the tides is pretty slim. The correlation between tidal timing and the position of the Moon was noted as early as the fourth century BC. High tides occur when the Moon is either high overhead or on the opposite side of the Earth. Thus, there are two periods of tidal oscillation per lunar orbit. And the intensity of the high or low tide, the amplitude, is correlated to the phase of the Moon. The highest tides are when the Moon is full or new.

Correlation does not prove cause, so it would be hasty to credit the Moon with causing the tides. The double oscillation per lunar orbit was particularly perplexing. Early explanations relied on a subtle heating of the water and the resulting expansion. But this would give the Sun a role at least as important as the Moon. And it doesn't account for the high tide that occurs when the Moon is on the opposite side of the Earth. Other explanations cited a magnetic-like effect from the Moon. Kepler favored this idea, but Galileo found it too mystical, this unseen action at a distance. He would only be satisfied with a mechanical cause acting directly on the water. He found it while commuting between Padua and Venice.

There are particular challenges to building and maintaining a city like Venice, set in a salty marsh and braided with canals. Where to bury the dead is one. Fresh water is another. You can't dig a well, so water must be imported from inland. Water for Venice was delivered in open barges, brought down the Brenta river. The barges also carried paying passengers, and this was a common way to get into the city. Galileo rode them frequently. He noticed that when a barge experienced an abrupt loss of speed, as when striking a dock or running up on a sandbar, the normally placid water would slosh up the front end of the container and down at the back, and then the flow reversed, rising up at the stern and down at the bow. This motivated experiments in Galileo's workshop with movable tanks of water. The results showed that periodic rising and falling of the water level on the edges of the container can be caused by periodically changing the speed of the container itself.

Now apply this to the ocean. Each ocean or sea is a huge open container of water. Galileo reasoned that the tides are caused by the container changing speed, just as in the barge. The changing speed is a result of the Earth having two motions, daily rotation and yearly orbit around the Sun. The axes of these two circular motions are more-or-less aligned and are in the same rotational direction. At a point on the Earth that is momentarily furthest from the Sun, that is, at midnight, the rotational motion and orbital motion are in the same direction. They add together. But 12 hours later, when that same point has come around to be at its closest to the Sun, at noon, the rotational speed is in the opposite direction from the orbital. The one speed is subtracted from the other. In other words, by Galileo's argument, that point, like every point on the Earth, speeds up at midnight and slows down at noon. Thus, each container of water that is an ocean or sea basin periodically speeds up and slows down. As on the barge, this causes the water to rise up on one side and recede on the other. These are the tides.

Galileo acknowledged some problems with his explanation of what causes the tides. The frequency of the lurching-barge model of the tides would be once a day, one high tide and one low every 24 hours. But the reality is roughly twice a day. He dealt with this by pointing out that the characteristics of water sloshing back and forth in a container are determined by the shape of the container itself. The barge gets only one push, after which the timing and height of periodic rising water depends on the size, depth, and shape of the boat. The oceans are no different. It's the extent and complicated shape of the ocean floor that accounts for the twice-a-day timing of the tides. The uneven motion of the Earth is just the basic driving force. It's like pushing a child on a swing—and this is my analogy, not Galileo's. Some force is necessary to get it going and sustain the oscillation, but the properties of the system, in this case the length of the swing, determine the frequency. You don't have to push every time the swing comes back; every other time will do. The driving frequency is half the frequency of oscillation. That's how it works with the tides.

As with the swing, the oceans require some outside force to maintain the ebb and flow. Galileo saw only one possibility, "if the terrestrial globe were immovable, the ebb and flow of the oceans could not occur naturally." In logical terms, the rotation of the Earth is necessary to cause the tides. It is also sufficient,

> when we confer upon the globe the movements just assigned to it, the seas are necessarily subjected to an ebb and flow agreeing in all respects with what is to be observed in them.

This is an unusually tight connection between a cause and effect. It's not just that the two motions of the Earth *would* cause the tides; it's the *only* thing that can. Thus, the fact of the tides is proof that the Earth rotates and orbits the Sun.

Galileo overstated the success of this explanation when he claimed that the predicted ebb and flow of the oceans are "agreeing in all respects with what is to be observed in them." There is no role for the Moon in Galileo's account. To ignore the correlation between the timing of the tides and the position of the Moon is to cherry-pick from the evidence in a suspiciously unscientific way. To avoid what he considered an occult influence by the Moon, Galileo simply ignored the Moon altogether.

The *ad hoc* addition to Galileo's theory of the tides, the ocean-basin determination of the frequency, does not address the much more fundamental flaw that Galileo seemed to have missed. The Earth is a rigid sphere, so it is implausible to think of one part of it speeding up while, at the same time, another part is slowing down. Physics routinely deals in implausible and counterintuitive ideas. It's business sometimes seems to be to challenge common sense. But in this case, intuition is right. It makes no sense to say that one part of the globe speeds up or slows down. The comparison to barges, where the whole vessel changes speed is illegitimate.

The real mistake in Galileo's explanation of the tides is that it violates his own principle of relativity. The hypothesized rotation of the Earth has a constant speed, as does the revolution around the Sun. The sum of these two uniform motions is a uniform motion. There is no speeding up or slowing down when the two motions are put together.

That's the principled case against Galileo's theory, with a glib appeal to uniformity. It may be too abstract, so here is a more intuitive explanation that takes us back to the tower argument. It's the uniform rotational speed of the Earth, and the steady speed of the tower, that allows the dropped stone to follow along and fall to the foot of the tower. If the Earth and tower were speeding up or slowing down, as required in Galileo's explanation of the tides, the stone would fall behind when the tower was speeding up or fall forward when the tower was slowing down. That's the inertia Galileo introduced, and the core property of the principle of relativity. If the tower argument is effective in showing that the straight fall does not disprove the rotation of the Earth, it is equally effective in showing that the tides do not prove that the Earth rotates.

Galileo had warned us, "whatever motion comes to be attributed to the earth must necessarily remain imperceptible to us and as if nonexistent, so long as we look only at terrestrial objects." This includes the tides.

It's fair to say that the evidence that the Earths rotates was inconclusive when Galileo stood before the Inquisition. This is including his own telescopic observations and his contributions to interpreting phenomena both celestial and terrestrial. Disagreement over the hypothesis would not have been unreasonable. In other words, was it reasonable to believe that the Earth rotates? Yes. Was it reasonable to doubt that the Earth rotates? Yes. Was it reasonable to ban or censor the books that presented evidence and argued for the rotation of the Earth. No. Was it reasonable to imprison or burn advocates of the Copernican model that included rotation and put the Sun at (or near) the center of things? Again, no.

CHAPTER 9
New Astronomy and the Great Magnet

Hence the entire terrestrial globe, with all its appurtenances,
revolves placidly and meets no resistance.
—William Gilbert

Galileo is a hero of science, not only for being on the right side of a dangerous scientific debate, but for outlining some of the most important activities and standards we call the scientific method. He emphasized the importance of securing the link between theory and evidence. He offered no explicit treatise on scientific method, and his scattered comments present a mixture of reliance on observation and logic, so the most effective way to understand his method is to follow what he does with the evidence. But near the end of the *Dialogue*, almost as an aside, he gets to the crux. "The method of investigation in natural science is to observe effects and figure out the cause." He considered the tides as an observed effect and figured out the cause. Effects are evidence, and the link to theory is in the causal connection.

Astronomy had never been concerned with causes. The celestial world is the epitome of stability, with eternal objects and uniform movements that continue without interruption or change. Stability, it seemed, needs no cause. Natural motion is innate, simply the way things are. Only the violent motion on the Earth requires an active influence, and that was the purview of physics, not astronomy. In light of Galileo's brief comment on the method of scientific investigation, the question was whether the cause-and-effect connection can be part of a science of astronomy. More fundamentally, did it make sense to ask about causes of celestial phenomena?

What requires a causal explanation, and what does not, may depend on the fundamentals of the conceptual and theoretical framework that directs the inquiry. Sometimes, stability does require a cause. Aristotle explained the spherical shape of the Earth and its stability at the center of the universe by citing the natural tendency of earth, the element, to move toward the central point in space. Falling in toward the center caused the formation of a sphere and continues to hold the Earth in place. In the same Aristotelian cosmology there was no corresponding need to explain the shape or rotation of the heavenly spheres. But with the loss of real solid spheres, the planets lost both their tether and means of locomotion. That introduced the need for real celestial causes, and invited the application of physics to astronomy.

Physics had been used for millennia to interpret the terrestrial evidence about the rotation of the Earth. To extend the application into the sky would provide a more comprehensive and more coherent theoretical network. This kind of unification is the sign of progress in science. As more things fit together, more things make sense. It also raises the standards for testing ideas, both new and old. Celestial science will have to be not only internally consistent in how it describes planetary and stellar motion, it will be held accountable to a dynamic explanation of the motion. There will be stricter constraints on the theory. And the dynamics that applies both on the Earth and above will be responsible for interpreting the evidence of the Earth's rotation.

Galileo began the union of terrestrial and celestial sciences by bringing the natural motion of the heavenly spheres, the uncaused movement in a circle, down to Earth, allowing the dropped stone to follow along, with no external force, as the ground rotates underneath. Natural motion is the same above as below. Johannes Kepler, Galileo's German contemporary, went further in bringing physics to the heavens, but by the opposite assumption; it's not that things like the stone can have two natural motions, but that no object whatsoever has any natural motion. All motion requires an active cause; being at rest is the only natural state of things. This applies to everything in the universe, on the Earth or in the sky. At the center sits the stationary Sun, naturally at rest. Kepler, like Galileo, favored the new heliocentric cosmological system with the planets, including the Earth, in solar orbit. Beyond the solar system are the fixed stars on a surrounding celestial sphere that never moves.

Kepler was a gifted mathematician; he was also something of a mystic. Since all the planets are in motion and this includes the Earth in both its rotation and revolution around the Sun, there had to be an interaction of some kind to compel them to move. With no established physics to cover such a novel idea, Kepler described the interaction as a "virtue" shared between one object and another, between the Sun and a planet. Each planet, he explained, is more than just an inanimate lump; it has a spirit and something like a mind. This is what attracts them to the Sun, and what motivates them to continue moving in orbit. It means there has to be a real object at the center of orbit to share the virtue. Planets wouldn't orbit an empty point in space, as in the Ptolemaic model with epicycles centered on unoccupied points on the deferent. Even the Copernican model put the center of planetary orbits at a moving point a small distance from the real Sun, again requiring the planet to be guided by nothing real.

Kepler's new dynamics would require revision to the Copernican cosmological model, starting with the position and role of the Sun. Planetary orbits need real anchors and guidance.

> A mathematical point, whether or not it is the centre of the world, can neither affect the motion of heavy bodies nor act as an object toward which they tend.

Orbit requires a real cause, as does rotation. Kepler published a detailed description of the planetary system that accommodates these dynamic needs in 1609. The book is usually known simply as *The New Astronomy*, and that is an accurate summary. But the full title is telling, *The New Astronomy: Based on Causes, or Celestial Physics, Brought Out by a Commentary on the Motion of the Planet Mars*. This is the first publication of celestial physics, what we now call astrophysics, and it's based on causes, the requirement that the celestial kinematics conform to the principles of universal dynamics. Contrast this with Copernicus, *On the Revolutions of the Heavenly Spheres*, which is strictly descriptive.

Before there was *The New Astronomy*, there was *Mysterium Cosmographicum*, the *Cosmic Mystery*. Kepler did not hide his predisposition to the mystical. He may have learned it from his mother, an herbalist and healer. At one point in her professional life she was accused of witchcraft. Kepler's father had little influence on his son, mystical or mathematical. A mercenary soldier, he abandoned the family when Johannes was 5. Like Copernicus and Tycho, Kepler was to grown up without a father in his life.

Johannes Kepler was born in Weil der Stadt, Germany, at 2:30 in the afternoon on December 27, 1571. The exact timing is by Kepler's own report, motivated by his keen interest in astrology. He can account for his conception with equal precision, 4:37 in the morning of May 16, 1571. Do the math and it reveals that the birth was significantly premature. A frail baby, he lived a life of chronic ill-health and weakness. The Keplers were not wealthy, but Johannes' sharp, mathematical mind won him a scholarship to Tübingen University when he was 18. The plan was to study for the Lutheran clergy, but, as happens, a love and aptitude for astronomy motivated a change of course. He left the university without a degree, taking a job in Graz, Austria as an instructor of mathematics and astronomy. Kepler wasn't a good teacher. There were just a few students in the classes of his first year at the school, but that's no way to evaluate someone's teaching, too soon to have a reputation, good or bad. In his second year he had no students at all. That's a bad sign.

It was while preparing an astronomy lecture in Graz, or perhaps during the lecture itself, that Kepler came to a realization about the structure of the universe. It included doubts about the details of the Copernican planetary model, and some fundamental ideas on how to fix it. With a reverence for the Sun reminiscent of Pythagoreans, and a tidy sense of harmony and coherence that may have been nurtured by his Lutheran training, Kepler looked for a unifying correlation between the sizes of planetary orbits and the length of time for their trips around the Sun.

This became the *Cosmic Mystery*. It also led to correspondence with Tycho Brahe, just when the great Dane was moving into his new facility in Prague. In 1600, Kepler got a job with Tycho, putting his mathematical and organizational skills to work on the collection of astronomical data. He was not much help in the physical task of observation, as an episode of smallpox as a child had left him with compromised eyesight. Tycho died suddenly in 1601, leaving the directorship of the observatory to Kepler, at one third the salary. More important, at least for the development of astronomy, Kepler inherited all the data.

It was during his 11-year tenure in Prague that Kepler wrote *The New Astronomy*. He was also assigned the duties as imperial mathematician and astrologer to Rudolf II, the Holy Roman Emperor, King of Hungary and Croatia, King of Bohemia, and Archduke of Austria, an oddball monarch whose incompetence is often cited as the cause of the Thirty Years War. Kepler eventually suffered the collateral damage of royal upheaval and was forced to leave Prague in 1611, moving back to Austria to teach. Moved around by the vagaries of a religious war and the demand for his astrological advice, Kepler eventually landed in Regensburg. He died in 1630, buried in a grave now lost in the destruction of the war.

The New Astronomy was published in 1609, the same year Galileo first looked at the sky through a telescope. The book is not directly about the evidence for the rotation of the Earth, but its celestial physics builds the planetary system in which the Earth must spin to save the astronomical phenomena. And the requirement for real causes of celestial motion will ultimately come down to orbits driven by the rotation of the central object. The revolutions of planets and the Moon will provide evidence of rotation of both the Sun and the Earth, respectively.

In Kepler's celestial model, the Sun is king. It is the cause of planetary orbits and it sits stationary, not at the center of the orbit but at the hearth, or, in Latin, the *focus*. We have appropriated the term to refer to each of the two points that anchor an ellipse, the precisely defined geometric shape of an elongated circle. Kepler put the Sun at one focus of the elliptical orbit of each planet. Following the trajectory of an ellipse put the planet on a single, smooth path, eliminating all of the epicycles of the Copernican model. Without epicycles, and with the Sun the actual focus of the orbit, Kepler's planetary system requires no empty points in space to do the work of defining or constraining the orbits. It's all up to the Sun, the real Sun. There is the unoccupied second focus of the ellipse, but it's just a geometric artifact, playing no role in celestial affairs.

With the planets on elliptical orbits, Kepler initiated a complete break from the ancient cosmology. Not only was the symmetry of perfect circles missing, but the speed of each planet had to be different along various stages of the orbit, faster near the Sun and slower at greater distance. Such irregular and variable speed would require an ongoing and equally irregular force. Astronomy without perfect circles may have been too radical for Galileo; he all but ignored Kepler's contribution. There is no mention of the new astronomy in Galileo's *Dialogue*, where everything spins and orbits at constant speed and unchanging distance from the center.

The new astronomy required a new physics. Radically different kinematics, the elliptical orbits, needed radically different dynamics, forces in the heavens. The old astronomy, Copernicus and Galileo included, needed no forces at all. In this new system, the planets are moved by what Kepler called an *anima motrix*, an animated or spiritual motive. The details are vague, but the Sun emits both light and invisible, magnet-like rays, the latter able to push and pull a planet to move it around. The rays can pull the planet in close during part of its orbit, and push it back out on the more distant part of the ellipse. And if the Sun rotates, the rays rotate as well, pushing the planets around like spokes on a wheel. Just as the light from the Sun gets fainter at greater distance, the force of *anima motrix* must decrease for planets farther away from the focus. This explains why the more distant planets move more slowly than the inner planets, and in fact there is a regular correlation between distance from the Sun and orbital speed.

Every object, whether on Earth or in the sky, has an innate resistance to movement, according to Kepler's physics. This is what makes standing still the natural condition, and any form of motion unnatural. Some things resist more than others, and consequently respond to a force such as the *anima motrix* with less speed. This further contributes to the sluggish orbits of the large planets like Jupiter and Saturn, compared to the swifter Mercury and Venus.

Kepler had predicted the rotation of the Sun before there was any direct observation of such a phenomenon. It's a requirement of his celestial dynamics, the driving force at the hub of the solar system. It was a bold conjecture, just the sort of thing that scientists value as a way to test a hypothesis. But when Galileo used his telescope to find evidence that the Sun does indeed rotate, and in the same direction as the orbiting planets, there was no celebration or talk of confirmation of the new theory of celestial physics. Galileo continued to ignore both the elliptical orbits of Kepler's kinematics and the *anima motrix* of Kepler's dynamics. The mysterious magnet-like force offended Galileo's sense of a mechanical universe in which all interactions happened by contact. He saw no possibility of a causal interaction between separated objects, and there was no place for such action at a distance in his physics. Kepler eventually revised the terminology to describe the power of the Sun to move a planet, calling it *vis motrix*, simply a moving force rather than a spirited animation. But the mystery remained, and Galileo was unmoved.

The celestial physics must apply to what moves on the Earth as well, and to the rotation of the Earth. The physics is universal. A rotating Earth would have a magnet-like force just like the Sun, and it could also push things around in the direction of its rotation. But it's not just magnet-like; it can be an actual magnet. In 1600, William Gilbert published a treatise on magnetism that claimed the Earth itself is a huge magnet, with magnetic poles lined up along the same axis as the planet rotates. For Kepler, it was a natural extrapolation: the rotating Earth causes the orbit of the Moon. Since the Moon is some distance from the Earth, the strength of the interaction is weakened and the speed of the Moon is slow. That's why the orbital period is much longer than the rotational period of 24 hours, that and the fact that the Moon, like everything else, has a natural tendency to resist moving. If the Moon was closer or lighter, or both, it would be easier to push around and it would orbit with greater speed.

The evidence for something unobserved is most often found in its effects. A flu virus causes a fever, thus evidence of the flu is in its effect, the fever. By Kepler's dynamic reasoning, the rotation of the Earth causes the orbit of the Moon. Thus, the plainly observable and never controversial orbit of the Moon is evidence that the Earth rotates. There were some loose ends, the unspecified parameters and the mystery of interaction between objects separated by great distance. Kepler, though a skilled mathematician, provided no precise description of the strength of the force and how exactly it depends on the distance between things. And there was no clear measure of a body's innate resistance to movement. The details of celestial physics were vague.

Despite the missing pieces, Kepler applied the same mechanism to respond to the relentless argument against rotation that cites the trajectory of objects dropped from high places and projectiles tossed into the air. If the Earth is rotating, these things would be left behind, falling a significant and measurable distance to the west—the tower argument. Kepler had an easy explanation for why, even on a rotating Earth, the falling stone follows along with the moving ground and falls exactly to the foot of the tower. Just as the rotating Earth drives the Moon around in its orbit, the Earth carries the stone around as it rotates. A stone is much lighter than the Moon, so its resistance to motion is much less. And a stone is much closer to the Earth than is the Moon, essentially at the surface of the Earth, so the rotational driving force is at full strength. Put these two factors together and they explain why the stone, like all projectiles, follows around with the rotating Earth at the same speed of rotation. It's the *vis motrix* of the Earth's rotation that keeps the falling stone from lagging behind.

This explanation of the trajectory of a falling stone works to reconcile the evidence with the hypothesis that the Earth rotates, but it does not hold up to the analogous experiment on a moving ship. Galileo dealt with the tower argument by pointing out that a stone dropped from the mast of a ship falls to the base of the mast whether the ship is moving or not. No on-the-object-itself experiment can determine whether the object is moving or not. This is an early version of the idea of relativity. Galileo argued that it's because no force is needed to maintain the horizontal speed of the stone as it falls. Everything in the system follows along naturally and at the same speed, so nothing about the behavior of one component of the system will reveal a systemic motion. Kepler required an active force for any motion in any direction at all times. If a ship is sailing north while the stone is dropped from the top of the mast, the rotating Earth will pull the falling stone along with it to the east, but nothing pulls it along with the ship as it travels north. According to Kepler's dynamics, the stone would fall to the back of the ship, some distance south of the mast. This, of course, is not what happens, although the distance would be very small unless the ship was moving very fast. To use our modern terminology, Kepler's theory about what moves projectiles, and what moves the Moon and planets, is not relativistic. It doesn't apply uniformly to all reference frames of motion.

From our perspective, Johannes Kepler got the description of the solar system, the kinematics of planetary orbits and the rotation of the Earth, exactly right. He got the dynamics, the account of what causes and sustains the motion of planets and ordinary things like a falling stone, very wrong. He accurately described the trajectory of the stone dropped from the top of a tall tower, but for the wrong reason. This is clear warning that kinematics, observations of where things are and how they move, does not uniquely determine dynamics, explanations in terms of unobserved natures and interactions. This is important to remember, since the evidence for the rotation of the Earth is generally interpreted with the guidance of a theory of dynamics, a theory always vulnerable to change.

Kepler got the inspiration for his dynamic theory from William Gilbert's work on magnetism. Gilbert is often credited with the first systematic, empirically grounded study of magnetic phenomena. The results were presented in *On the Magnet and Magnetic Bodies, and on the Grand Magnet the Earth*, usually referred to simply as *De Magnete*. He did the work in his spare time since his real occupation was medicine. Cambridge-educated, wealthy, and without the distractions of wife or children, Gilbert made a notable career as a physician. He was the appointed doctor to Queen Elizabeth I, and when she died in 1603, Gilbert's new patient was King James VI. Little is known about his personal life, since any relevant documents were lost in the Great Fire of London in 1666.

We do know that Gilbert endorsed the new heliocentric cosmology from the start, referring to "Copernicus, restorer of astronomy." His interest was on the terrestrial aspects of the Copernican theory, the rotation of the Earth, with little to say about the Earth's revolution around the Sun or the orbits of other planets. Gilbert was not an astronomer. Nor did he suffer politely the advocates of the old world system, and much of his refutation of the Aristotelian ideas was rooted in name-calling. On the notion that the celestial sphere of stars revolves around the Earth once a day, for example, "Surely that is superstition, a philosophic fable, now believed only by simpletons and the unlearned … while the importunate rabble of philosophers egged them on."

The stars do not orbit the Earth; the Earth rotates beneath the stars. This, for Gilbert, was neither superstition nor fable; it was confirmed by good evidence. It starts with the idea that the Earth itself is a grand magnet, demonstrated in the evident fact that compasses spontaneously and consistently line up along a longitude and point north. Earlier explanations of the behavior of the compass or other small magnets free to pivot described the orientation by reference to the stars and the celestial sphere. A compass points to the north star. Gilbert shifted the phenomenon of magnetism from the heavens to the Earth itself. And this provided an explanation of the Earth's rotation, since, "… all magnetic bodies (when fitly arranged) are borne round in a circle." The "fitly arranged" simply means free to spin around, like the needle of a compass or the Earth itself. Thus, the inherent magnetism of the Earth both drives the rotation and keeps the axis of rotation aligned with the stars.

Strictly speaking, Gilbert's appeal to magnetism is not evidence that the Earth rotates; it's an explanation of the rotation. One must already believe that the phenomenon is real to ask why it happens. If anything, rotation would be the evidence of magnetism, since rotation is cited as the effect. But showing that a presumption of the Earth in rotation fits consistently into a larger physical theory, now including magnetism, adds to the reason to believe the whole system of ideas may be true. Gilbert was confident in the conclusion. "From these arguments, therefore, we infer, not with mere probability, but with certainty, the diurnal rotations of the earth."

Like all proponents of a rotating Earth, Gilbert was obligated to comment on the tower argument. His explanation of the falling stone following the moving tower was not new, nor was it ever very precise. Again, his impatience with the old world system was evident. On the reason that the atmosphere and clouds and projectiles are not left behind the rotating Earth, "all the circumfused effluences and all heavy bodies therein, however shot thereinto, advance simultaneously and uniformly with the earth because of the general coherence." It's not clear what he meant by "general coherence." Aristotelian scholars claim the dropped stone would fall some distance to the west on a rotating Earth, "but these are old-wives' imaginings and ravings of philosophasters ..."

In the early seventeenth century, Gilbert was the undisputed master of magnetic studies, but experiments with magnetism were challenging and delicate, and the understanding of the phenomena was still vague and imprecise. Magnetism is a very weak effect, at least with naturally occurring loadstones and the magnetic field of the Earth at its surface. Experiments are susceptible to disturbances that overwhelm or distort the results, and the interaction itself, what we now describe in terms of a force field, is invisible. It was mysterious then, perhaps a little less so now. Given the challenges, it's to be expected that there were disagreements about the nature of magnetism.

While Gilbert cited magnetism as the force that moved the Earth, Athanasius Kircher claimed in 1641 that it was the force that held the Earth stationary and kept it in its place, that is, at the center of the universe. Again, the analysis of the interaction was vague, and Kircher argued that the Earth itself could not be a magnet, otherwise anything composed of iron would be drawn to the ground. It would be impossible to use iron tools. He also pointed out that experiments with spinning magnets in the laboratory never caused nearby objects to follow the rotation by orbiting the magnet. Kepler's theory about the orbit of the Moon simply did not match any analogous evidence on the Earth.

The more direct refutation of Gilbert came from Jacques Grandami. He described his own experiment in which a floating loadstone, left free to move in any way, did not rotate. This he cited as proof that the Earth does not rotate. On the assumption that the Earth is a magnet, and the experimental demonstration that the loadstone, fitly arranged, does not rotate, Grandami drew the clear conclusion that the Earth does not rotate either. Magnetism does not move the Earth; it holds the Earth in place. Grandami wrote in 1648, "The goal of magnetic virtue is the good and quiet of the earth."

The laboratory experiment cited by Grandami depends on an analogy. The Earth is like the loadstone, so what is known by observation about the loadstone can be inferred to be true of the Earth. The physics of magnetism, the dynamics, is being used to interpret the astronomical data of the diurnal motion of stars and planets, the kinematics. As always, the evidence, whether it is for or against the theory that the Earth rotates, must be interpreted in the context of some dynamic theory. Since rotation itself, at least the absolute rotation that is the issue in the seventeenth century, is not observable, the evidence will be circumstantial.

Disagreement on whether or not a spherical magnet would spontaneously rotate took on a public and practical aspect with the word of a so-called magnetic clock. In 1634, Sylvestre di Pietra-Sancta published a book on symbolic artifacts in which he described and pictured an invention by a fellow Jesuit, Francis Linus. A ball suspended in water within a glass sphere was seen to rotate, spontaneously, once around in 24 hours. In Pietra-Sancta's telling, the orb by an arcane force and as if by a certain love strives after the conversion of the sky from east to west and is driven around altogether in the space of 24 hours.

With the 24 hours of a day marked on the outside of the glass, the device functioned as a clock, with the arcane force credited to magnetism.

Nicholas Claude Fabri de Peiresc, an amateur astronomer and advocate of the Copernican model of the solar system, saw this magnetic clock as a way to prove that the Earth rotates. He wrote to Galileo, then under house arrest after condemnation by the Roman Inquisition, suggesting that the clock might be his ticket to exoneration. And the truth shall set you free. But Galileo already knew the truth about the clock, that it was a hoax. He claimed to have not only heard about the magnetic clock but to have built one himself. It only worked because there was a hidden mechanism beneath the table, a water-driven clock that turned a magnet to drive the spinning orb above. Magnetism wasn't the driving force; it was only part of the linkage between the waterwheel and the orb.

Athanasius Kircher had also heard about the magnetic clock and, like Galileo, constructed his own. He published a diagram of the hidden mechanism required to keep the orb spinning, with an argument that this artifice proves that a magnetic sphere will not rotate on its own, neither the one displayed in the glass ball of the clock nor the alleged grand magnet the Earth. The logic in this is a bit sketchy. Building a clock and showing that the magnetic sphere *can* be induced to rotates by means of an external drive does not prove that it will rotate *only* when there is the added influence. The water-driven clockwork is sufficient to turn the orb, but it may not be necessary. The experiment was ambiguous, and consequently the evidence about the Earth's rotation was inconclusive.

None of the magnetic clocks in this debate have survived. There is a modern replica in the library at Stanford University, built in 2001, based on the design illustrated by Kircher. But it doesn't have the water-clock underneath to drive the orb; it runs by an electric motor.

The dynamics of magnetism as an interpretive support of evidence for, or against, the rotation of the Earth was not much help in the seventeenth century. It was a mysterious force, difficult to measure and challenging to understand, both in its source and its effects. Consequently, evidence of rotation still appealed to the larger cosmological system and the role that diurnal rotation played in a coherent structure of stars, planets, Sun, and Moon. There was no evidence specific to rotation, only that it was a part of a larger system that consistently and elegantly saved the phenomena.

CHAPTER 10

Rotational Dynamics and Absolute Space

I will communicate to you a fancy of my own about discovering the Earth's diurnal motion.
—**Isaac Newton**

As long as we are stuck on the Earth, we cannot directly observe its motion. And until there is clarity on the local effects of rotational motion, there can be no clear evidence of the Earth's rotation. Most physics students will tell you that understanding the details of circular motion, the causes and effects and even just the exact description of rotation, often demands ignoring your intuitions and giving in to the mathematics. Richard Feynman, in the chapter on "Rotations in Space" in the textbook transcription of his lectures to Cal Tech freshmen, warned the students that, "there are circumstances in which mathematics will produce results which *no one* has really been able to understand in any direct fashion." But until the middle of the seventeenth century, there was no mathematics of rotational dynamics, nothing to calculate. Intuitions ruled the interpretations of evidence, and intuitions varied.

Well before there was a science of dynamics or a precise mathematics of motion there was explicit recognition of a fundamental difference between straight and curved trajectories. Aristotle allowed both the straight line and the circle as representations of natural motion; once begun they required no sustaining cause. But he kept them separated between heaven and Earth—circles in the sky, straight lines near the ground. Galileo continued the Aristotelian distinction between straight and circular as two different forms of natural motion, but allowed them to act together in the flight of a projectile on the Earth. Kepler required a cause for anything that moved, but distinguished the force that pulled objects toward each other on a straight line from the sweeping force that moved things around on the elliptical curve. There was disagreement about causes, and disarray in the mathematical description of the dynamics, but there was also universal agreement that curved motion, including rotation, needed to be treated differently from straight line. Clarifying the distinctive and beguiling properties of circular motion would be pivotal to interpreting the evidence that the Earth rotates.

The increasingly precise measurements of celestial phenomena were about circular motion, but the results had not been decisive in proving that the Earth rotates. At best the data from the sky may have favored one cosmological system over the other, by criteria of simplicity or metaphysical principles about the importance of the Sun or the requirement of celestial circles, but no data from the sky were specifically about the Earth's rotation. On the other hand, the measurements of terrestrial phenomena that were meant to demonstrate the Earth's stability or its rotation had not been detailed or careful. They referred only to our common experiences of tossing stones and enjoying the calm atmosphere. The tower argument never needed precise measurement, since by general agreement the effect in question, the stone falling some distance to the west, would be enormous, on the order of a kilometer. It would be obvious, if it were real. The question had been whether it would happen at all if the Earth is rotating. Terrestrial evidence is about the Earth itself, but it must be interpreted through the understanding of rotational dynamics, the cause and the effects of the rotation.

Any acquaintance with the history of physics will have you anticipating Isaac Newton at this point, to put the mechanics together and sort out the mathematics of rotational dynamics. He did, in 1687, with the publication of the *Mathematical Principles of Natural Philosophy*. The *Principia*, as it's usually called, is the operating manual for the universe, covering both the celestial and terrestrial phenomena under a single system of mechanical laws. It's so tidy and effective that we often overlook the struggle to put it together. Newton had help, although he was not altogether gracious in acknowledging it. He had help in particular from fellow Englishman Robert Hooke.

Robert Hooke was a little older than Newton. He was born in 1635 and died in 1703. He was a man of diverse talents and interests, including mathematics, architecture, chemistry, astronomy, and microscopy. He introduced the word "cell" to describe the structural units of organisms. He worked with Christopher Wren to design buildings in the reconstruction of London after the fire of 1666. And he wrote a book entitled, *Attempt to Prove the Motion of the Earth*. This was in 1674. The attempt would rely on detailed terrestrial evidence, and Hooke recognized the need to understand the dynamics of rotation to make sense of what he was measuring. He also hoped the discussion of rotation would be a way to reconcile with Isaac Newton, whose brittle ego had been injured over an earlier dispute about optics and the composition of colors.

Hooke reached out with a letter to Newton in 1679. He discussed the work to be done to use Kepler's careful description of planetary orbits to derive equally precise laws of the forces holding the system together. Newton replied, and indicated some interest in the project, but said he had a more immediate task. "I will communicate to you a fancy of my own about discovering the Earth's diurnal motion." This was 1679, more than a 100 years after the the publication of *On the Revolutions of the Heavenly Spheres*, and still two of the most distinguished scientists were hoping to find the evidence of one of the key components of the Copernican system, the rotation of the Earth.

Isaac Newton was born on Christmas day, 1642, in Woolsthorpe, England. His father had died before the birth, and within 2 years his mother remarried, leaving the frail infant to be raised by his grandparents. The boy, not surprisingly, hated his distant stepfather. But it only lasted 9 years before his mother's second husband died, and Newton went home to work the family farm. He inherited from his stepfather a small notebook, and the promising mathematician kept careful record of his developing ideas about nature. He called it his Waste Book.

In 1661, Newton enrolled at Cambridge University. Classes in the sciences, or natural philosophy, as it was then, were still conducted under the canon of Aristotelian ideas. Newton developed as a mathematician and received his degree in 1665. An outbreak of bubonic plague closed the school for 2 years, sending the new graduate home where he developed many of his foundational ideas. It's during the sequester that he produced a treatise on optics, describing light as a stream of tiny particles and explaining such phenomena as the focusing power of a lens and the dispersion of colors by a prism in terms of the speed of the particles in glass. This is the work that drew Hooke's attention and critique, the first of their many disagreements.

Newton returned to Cambridge in 1667 as a professor of mathematics. He was there for 32 years. He moved to London in 1696 and was appointed as warden of the mint, developing a reputation as an uncompromising guardian of monetary propriety. He died in London in 1727, never married, and leaving no indication of romance or relationship. And despite being born 30 km from the English channel and having figured out the cause of the tides, there is no record of Isaac Newton ever having seen the ocean.

The dispute with Hooke began over optics and moved on to celestial mechanics. They disagreed on which of them first proposed that the force holding the planets in orbit weakened with distance, specifically as the inverse of the square of the distance. But they agreed on the possibility of using earthbound experiments to demonstrate that the Earth rotates. Both of them, with Hooke's attempt to prove and Newton's fancy about discovering, returned to the ageless tower argument. They hoped that by dropping stones from great heights they would not just undermine Aristotelian arguments against the possibility of rotation but to reveal evidence that proved the Earth does rotate. It all depended on the laws of rotational dynamics.

Hooke laid the groundwork for distinguishing evidence of rotation from evidence of straight-line motion in three "suppositions" at the end of the *Attempt to Prove the Motion of the Earth*. They are remarkably like some of the foundational claims that show up later in Newton's *Principia*, and it is easy to see why Hooke thought his own contributions had been overlooked.

The first supposition sounds a lot like universal gravitation.

> First, That all Coelestial Bodies whatsoever, have an attraction or gravitating power towards their own Centers, whereby they attract not only their own parts, and keep them from flying from them, as we may observe the Earth to do, but that they do also attract all other Coelestial Bodies that are within the sphere of their activity ...

Thus, the same attraction that holds pieces of the Earth together and draws them to the ground also provides the tether to celestial objects in orbit.

The third supposition, and it makes sense to consider them out of order because the third is a follow-up of the first, explains that this attractive gravitating power must decrease with distance.

> That these attractive powers are so much the more powerful operating, by how much nearer the body wrought upon is to their own Centers.

It's not a quantitative law, and Hooke offered it as a challenge for Newton to derive the exact mathematical relation between the force and the separation between bodies. Hooke would later claim to have worked it out for himself, the inverse-square formula, before Newton went public in the *Principia*.

The second supposition is a law of inertia, and it's the one most important for the evidence of rotation.

> That all bodies whatsoever that are put into a direct and simple motion, will so continue to move forward in a straight line, till they are by some other effectual powers deflected and bent

into a Motion, describing a Circle, Ellipsis, or some other more Compounded Curve Line.

Natural motion is straight; it requires no sustaining force. Any curve, including orbit or rotation, requires a force. Newton subsequently made this mathematically precise in the second law, $F = ma$, in which the acceleration a is any change of speed or direction, that is, any deviation from the direct and simple straight line.

This is the crucial link between kinematics and dynamics, between how things move and the forces required to make them move. Newton put it all together in the *Principia*. He introduced gravity as the central force holding the planets in elliptical orbits and described the heliocentric system with exact mathematical expressions. It worked with the Sun at the center—that is, at a focus of the ellipse—but not with any model that had the Earth stationary at the center. The comprehensive and coherent new world system was now complete.

Planetary orbits need a central force that generates an ellipse. Similarly, a rotating sphere must have a force that holds things together. Add in Newton's third law of motion from the *Principia*, the one about every force being matched with an equal and opposite reaction. The tethering force, the one pulling things in toward the center of rotation is the centripetal force. It must generate an outward force of equal magnitude; this is a centrifugal force. An effect of rotation will be an outward pull. If this can be measured there will be evidence of the rotation.

The term "centrifugal" to describe the outward reaction to circular motion was introduced by Christian Huygens, a Dutchman. His *De Vi Centrifuga* was written in 1659, but not published until 1703. In the meantime, he circulated his analysis of centrifugal force and acceleration. The basic idea was certainly not new; Ptolemy had worried that on a rotating Earth, unsecured objects would be thrown off into space. Huygens made it precise by deriving the mathematical formula for the acceleration as a function of the speed of an object and the radius of its circular path. This allowed calculation of the magnitude of the effect on a rotating Earth, showing that even at the equator, where the speed was the greatest, the centrifugal acceleration, the throwing off effect, would be insignificant.

The effect would be insignificant, but not exactly zero. Very delicate measurement might reveal the centrifugal effects of the Earth's rotation. A meaningful test would require a specific prediction, and this is the task that Hooke attempted and that was Newton's fancy. For the first time there was the suggestion that small but detectable effects on the Earth itself could reveal its rotation.

At this point it is worth clarifying that it is absolute rotation of the Earth that Newton and Hooke were pursuing. Rotation in space itself, they were implying, is detectable. It's the sort of thing we feel for ourselves when spinning around or going around a fast corner. It's not spinning with respect to the Sun or spinning with respect to the stars, not solar curved or sidereal curved; it's simply curved in space. There is an implicit reference to an absolute space and absolute rotation. Newton made it explicit, describing the position and movement of things in an invisible universal container, space itself. There is a real distinction between one point or another in space, between moving through space or not, and between a straight or curved trajectory in space. So, there is a real question whether the Earth rotates or not. It's not simply rotation relative to the Sun or the stars; it's absolute rotation.

Straight-line motion, as in Hooke's second supposition, is "direct and simple"; It is of uniform speed with respect to the inherent grid of absolute space. It's real, but by itself undetectable. We can only see and measure this kind of motion if there is something else, some other visible object, to use as a reference. Only uniform motion relative to things is observable, even though uniform motion in absolute space is real. In this, Newton breaks from the strict dependence on empirical evidence that he professes on the occasions he discusses his scientific method, this and the idea of an invisible gravitational force that instantaneously influences bodies at great distance.

Nonuniform, accelerated motion is different. Rotation is in this category. Absolute rotation, rotation in space itself, is not only real without reference to any other objects, it is detectable without any reference to other things. It's locally detectable by characteristics of the rotating object itself. That's because of the need for a centripetal force to maintain the rotation, and, importantly, the equal and opposite centrifugal force. These rotational dynamics are the key to on-the-ground evidence of the Earth's rotation.

Newton provided a useful example to make the point about detecting the absolute rotation of an object. His thought-experiment with a rotating bucket has been the go-to prop for clarification on absolute space and rotation. Imagine a bucket half-full of water, suspended on a rope that is attached to the ceiling. The surface of the water is flat. Now, twist the rope around a few times so the bucket is set to spin on the vertical axis when it is released. At first, the bucket is spinning but the contained water is not, since it takes some time for the friction along the sides of the bucket to get things going. With the water still not rotating its surface is flat. But once the water begins to follow the spinning bucket around, once the water itself is rotating, it rises up the sides and the surface becomes concave.

When the water is rotating, its surface is concave. This is a real, observable, measurable property. When the water is not rotating, its surface is flat. It's not rotation relative to the bucket that counts, since the water is flat at the beginning, when the water and the bucket are both stationary, and also flat at step two, when the bucket rotates but the water doesn't, but then concave at the end, when water and bucket are again in synch. It's not rotation relative to the walls of the room that counts, since you can image just putting the whole room on a carousel and having it spin with the bucket. No, it's simply rotation, relative to absolute space. Newton continued the thought-experiment by describing two weights, joined by a rope, floating in an otherwise empty universe. There will be tension in the rope when the binary system rotates, but not otherwise. You could also image a water balloon in similarly empty circumstances. Rotation will cause the balloon to bulge at the equator.

This is good news if you fancy discovering the Earth's diurnal motion. It shows that an observable and measurable feature of the object itself, the change in its shape, can reveal the rotation of the Earth. Like the water balloon, the rotating Earth will bulge at the equator.

The Earth is solid, but not entirely. There's liquid both inside and out—the outer core is liquid, and so, of course, is the ocean—and even the solid layers, the mantle and crust, are not completely rigid. Early in the formation of the planet, when it was hotter and the solid parts were more malleable than they are now, rotation would affect the Earth like the water balloon. The formative bulge at the equator would then be frozen in place as the planet cooled and solidified. A rotating Earth will not be a perfect sphere; it will be, in the terms of geometry, an oblate spheroid.

Newton's own reasoning to connect rotation with an equatorial bulge was less direct but relied on less geological theory. The surface of the Earth is mostly water. Rotation will draw the oceans toward the outer edge of rotation, toward the equator. This would flood the tropics if the land near the equator didn't also bulge, rising as high as the centrifugal-risen sea level. This is in the *Principia*

> ... if our earth were not a little higher around the equator than at the poles, the seas would subside at the poles and, by ascending in the region of the equator, would flood everything there.

The effect of a rotating Earth is a bulging equator. Newton was pleased to point out that the prediction of equatorial bulge matched some existing celestial data. Jupiter was observed to rotate and it is visibly oblate, bulging at its equator.

Newton, as was his wont, did the math. He calculated that the diameter of the Earth at the equator would be greater than the diameter at the pole by one part in 230. This facilitated comparison to evidence that had been recently presented showing that a pendulum clock ticks more slowly near the equator than near the north pole. The connection between the swinging pendulum and the shape of the Earth required some explanation. The period of a swinging pendulum depends on its length and on the strength of gravity. It will swing slowly on the Moon, where gravity is weak, and it will swing more slowly the longer it is. In 1672, Jean Richer took a pendulum clock to Cayenne Island, just 4° north of the equator. The clock had been calibrated in Paris to swing exactly once per second, but he found he had to shorten its length to keep the same timing on the equatorial island. This was not an easy adjustment to do properly, since the pendulum was the only kind of clock available. There was no standard one-second to compare to the pendulum that needed adjustment. Richer must have used a celestial reference, carefully counting the swings of the pendulum for an entire solar day. His results were credible enough to get the attention of both Newton and Giovanni Domenico Cassini, the director of the Paris Observatory and the man who had measured the oblate shape of Jupiter.

Cassini and Newton agreed that the shorter one-second pendulum at the equator indicated that the gravity is weaker there than in Paris. They disagreed on the reason. Cassini argued that weaker gravity was a result of there being less mass under the pendulum, less earth. The planet must be skinnier at the equator, not fatter. He concluded that the Earth is a prolate spheroid, like a Roma tomato. Newton, who knew much about gravity, said the weaker gravity was because the Cayenne pendulum was further from the center of the Earth than the pendulum in Paris. This indicates a bulging equator. His calculations of the rotational bulge and its effect on the period of a pendulum matched Richer's observations.

There are a lot of steps in this inference about rotation of the Earth: The rotation changes the shape which alters the strength of gravity that in turn affects the period of the pendulum. The complication introduced significant ambiguity in interpreting the connection between the period of a pendulum and the rotation of the Earth. Fortunately, there is a much more direct way to determine the shape of the Earth. The curvature and shape of an object can be measured while on the surface, without having to rely on any dynamic theory to filter the interpretation. Eratosthenes did it in antiquity, measuring the curvature of the surface and from that inferring the diameter of the sphere.

In 1736, the French Royal Academy of Sciences sponsored two geodetic expeditions, north and south, each to measure the ground length of 1° of latitude. If Newton was correct and the Earth bulges at the equator, the length would be longer in the south, near the equator, than in the north, near the pole. The geometry stretches to accommodate the bulge. The southern team was sent to Peru, where they encountered even more troubles and delays than you would expect in the tropics and Andes. They were gone for 10 years. Those who went north, to Lapland, were more fortunate, returning after only a year, with celebrated results.

The Lapland group, led by Pierre Louis Moreau de Maupertuis, did their work in the Tornio River Valley, Sweden, at 68° north latitude. It was no picnic for them either, as they suffered through the cold and darkness of winter and clouds of insects in the summer. Latitude was measured by the angle to the celestial north pole, and tracking along a north-south line, along a longitude. This forced the crew through rivers, over rough terrain, and through dense forest. They returned in 1737 with the data, and, not waiting for the return of the southern expedition, compared their measurements to what had already been done on the ground near Paris. The results were remarkable. Not only was the ground-length of latitude shorter in the north, near the pole, it was considerably shorter than predicted, indicating a flattened pole and elongated equator of a rotating Earth. Maupertuis was a hero. He still is, with monuments on the ground in Sweden at both ends of his survey.

The journey to Lapland was almost unnecessary, given the confidence among scientists by that time in both the Copernican cosmology and the Newtonian dynamics to go with it. They knew the Earth rotated, even without this evidence. Voltaire, a friend of Maupertius, teased the returning expedition leader

> *Vous avez confirmé dans les lieux pleins d'ennui Ce que Newton connut sans sortir de chez lui.* (You have confirmed, in dreary far-off lands, What Newton knew without ever leaving home.)

This may have been over-confidence. It's not uncommon in the history of science to credit an experiment as being well done just as long as it delivers confirming results. It's like certifying an election as fair, just as long as your candidate is the winner. By the time the southern measurement team returned from South America, Maupertuis' data were being questioned, suspected of being too good to be true. The results from Peru matched Newton's prediction of equatorial bulge more closely than those from Lapland, motivating a review of the earlier work in the north. Numerous mistakes were revealed, and it was just lucky that the errors accumulated on the side of shorter lengths, hence oblate spheroid. It could have gone the other way.

By the mid-eighteenth century, the rotation of the Earth was generally believed by the community of scientists. There was consensus. Equatorial bulge is real, and it is caused by centrifugal force and this is the result of rotation. The Aristotelian standard model had been replaced by the Keplerian solar system and its companion physics, Newtonian mechanics. Nonetheless, the geodetic data from Lapland needed to be corrected. The Swedish Academy of Science made plans to revisit the northern latitude, but it was not until 1801 that Jöns Svanberg returned to the Tornio Valley. He found Maupertuis' measurements to be off by 400 meters. His own results were consistent with Newton's predictions and the evidence from Peru.

So, here was evidence that the Earth rotates. It came not by looking up into the sky but down at the ground. The shape of the Earth indicated the motion of the Earth, and the connection was through the dynamic concept of centrifugal force. It's difficult, and probably unimportant, to put a name or a date on the discovery of this evidence, whether it was Richer with the pendulum clock in Cayenne, or Maupertuis with the mistaken measurements but the right conclusion, or Svanberg with the corrections or Charles-Marie de la Condamine who led the 10-year expedition to Peru. It doesn't really matter; the Earth bulges at the equator, and that is evidence that it rotates on the north-south axis.

The equatorial bulge was not what Hooke had in mind with his attempt to prove the motion of the Earth, nor did it factor in Newton's fancy about discovering the Earth's diurnal motion. They were both thinking about the tower argument. Aristotelians had argued for 2,000 years that the perpendicular fall of a dropped stone proved the Earth does not rotate. Copernicans insisted that the perpendicular fall proved nothing. Newton and Hooke both thought that the trajectory of the stone could prove the Earth *does* rotates, because the fall is in fact not exactly perpendicular.

Newton applied his understanding of rotational motion to the flight of the dropped stone. He concluded that if the Earth is rotating the stone won't fall straight down; it will fall just a little bit east of the tower, that is, in the same direction as the rotating ground, the opposite direction predicted by Aristotelians. In a letter to Hooke, he explained that the stone, "... will not descend the perpendicular, ... but outrunning the parts of the Earth will shoot forward to the east ..." At the top of the tower, the stone is moving eastward faster than the ground, simply because the tangential speed of rotation increases as the radius increases. It's simple geometry. With no horizontal force acting on the falling stone, it maintains its original horizontal speed, faster than the ground, and so outruns the tower. It's a very small effect, since the height of the tower is so much smaller than the radius of the Earth, and Newton provided no calculation of the distance. His fancy only suggested that it would be measurable if the tower were tall enough. He elaborated on the shape of the stone's trajectory, describing a spiral that, if extended into the Earth, would eventually strike the center of the planet.

Hooke replied. He agreed that the stone would fall ahead of the tower, a small distance to the east, but also with a small drift to the south. The trajectory, though, is not a spiral but an ellipse, exactly as a planet orbiting the Sun or the Moon orbiting the Earth. It would not hit the center. What is the stone, after all, but a tiny planet with initial tangential speed and under the influence of the central force of gravity? Hooke went on to tell Newton that he himself had done the experiment. The results seemed to confirm their prediction, but having dropped only three balls and found three different points of landing, he was unsure of his findings. All three fell to the southeast,

and that very considerably, the least being a quarter of an inch, but because they were not all the same I know not which was true. What the reason of the variation was I know not, whether the unequal spherical figure of the iron ball, or the motion of the air, for they were made without doors, or the insensible vibration of the ball suspended by the thread before it was cut.

Newton grudgingly admitted his error in thinking the trajectory would be a spiral. Having been caught making a mistake did not help his chilly regard for Hooke. But it did reaffirm his belief that a carefully controlled experiment could confirm the rotation of the Earth.

Hooke and Newton were not the first to suggest that the falling stone in the tower argument would fall to the east on a rotating Earth. Galileo had mentioned it in the *Dialogue* in 1632. His on-again off-again relationship with his own principle of relativity allowed him to reevaluate the trajectory of the falling stone from the perspective of an absolute reference frame. The stone has the circular motion initiated while being held at the top of the tower on a rotating Earth, but he reasoned that it must also be on course to strike the center of the Earth. This is the combination of the two natural motions, circular and vertical. He concluded,

> far from failing to follow the motion of the earth and necessarily falling behind, it would even go ahead of it, seeing that in its approach toward the earth the rotational motion would have to be made in ever smaller circles, so that if the same speed were conserved in it which is had within the orbit, it ought to run ahead of the whirling of the earth, as I said.

The stone ought to fall to the east of the tower, but by an undetectably small distance.

The inconsistency in Galileo's reasoning was spotted soon after publication, even by Galileo. But the idea of an eastern trajectory was in play. Pierre Fermat treated the situation as a strictly geometric exercise and in 1636 decided the trajectory was not the semicircle as proposed by Galileo but a spiral, anticipating Newton's first thoughts. The stone would fall east, but by a very little bit. Giambattista Riccioli, an outspoken critic of the heliocentric cosmology that put the Earth in rotation, followed fallacy with fallacy in 1651 by arguing that, since Galileo's analysis was flawed, his conclusion that the stone falls to the east is false. And, since the analysis was based on the Copernican model of the cosmos, that too must be wrong. Another Italian, Giovanni Borelli, wrote in 1668 that the stone would not have to hit the center of the Earth, but that its circular impetus, very much like the modern concept of angular momentum, must carry the falling object faster than the tower to strike the ground a very small distance east of the point directly below its release. He had trouble believing his own result, an early case of intuition at odds with the results of mathematical analysis of rotational motion. The disagreement was academic for Borelli, since any deviation from a perpendicular fall would be too small to measure.

Despite the confusion over the math, the possible eastward deviation seemed like a prediction worth testing. It's a very challenging experiment, because the effect is so small. It's not like the Aristotelian prediction of the stone falling to the west, where the distance would be over a kilometer when dropped from a typical Italian tower. Hooke had tried, and maybe seen a fraction of an inch deviation when dropping balls from a height of 30 feet (9 m). He recognized the sensitivity of the process, and how prone to perturbation the flight of the ball or stone. There are ambient air currents. The falling object creates its own turbulence. And the release of the falling object has to be a passive drop, without the slightest nudge or twist that would impart an extra horizontal motion. There is a trade-off between the height of the tower and the disturbing effects. The higher the drop, the more distance to the east is predicted, but also the more time for disruptive influences.

There were numerous attempts to measure the fall of a dropped ball with sufficient precision to test the eastward prediction and find evidence of the Earth's rotation. After more than a 100 years of confounding challenges, Giovanni Guglielmine delivered credible results in 1791. He dropped lead balls, one inch (2.5 cm) in diameter from the top of the Asinella tower in Bologna. It's a fall of 241 feet (75 m). Each ball was held by a thin string and observed with a microscope to determine when it was perfectly still. Only then was the ball released to fall onto a bed of wax on the ground. The average displacement from a vertical descent was three-quarters of an inch (about 2 cm). The vertical reference was determined by hanging a plumb-line from the point of release. This, too, was vulnerable to wind and weather, and it was only 6 months after the experiment that a reliably vertical line was dropped.

Subsequent work was similarly encouraging (for those who believed the Earth rotates), but equally uncertain. Sometimes there was a southern component to the fall, sometimes not. There was a need for both more reliable measurement and a better understanding of the forces at work. Testing a hypothesis requires an exact prediction. In 1802, two of the world's leading mathematicians were called in to calculate the predicted deviation from perpendicular in a planned drop down a 90-meter mineshaft in Schlebusch, Germany. Carl Friedrich Gauss was a young man of 24, but already an accomplished and respected mathematician. He had successfully determined the orbit of a recently discovered minor planet, Ceres, using the available data that covered only 3° of its arc across the sky. He would go on in his carrier to provide the foundational understanding of non-Euclidean geometry, the language of general relativity. Pierre Laplace was older. At 53 he had an almost encyclopedia of mathematical accomplishments, including finally putting to rest the worry that Newtonian mechanics described an unstable solar system in which a small bump of a planet would cause the whole thing to fly apart. Laplace did the math to show that planets would settle back into their orbits after small perturbations. There was no need of an intervening god, as Newton suggested, to regularly restore order. Laplace, sometimes called the French Newton, also made an early prediction of black holes, stars so massive that their gravitational force prevented light from escaping.

Laplace and Gauss independently made the same prediction of a 9-mm deviation to the east for the balls dropped in Schlebusch. The results from the mine were reported in 1803 to be 8.5 mm. Thus, with a reliable theory of rotational dynamics, and skilled mathematicians, did dropping balls down a mineshaft in Germany deliver evidence that the Earth rotates. A celebration is warranted, but only mindful of the sketchy logic involved. If the Earth rotates, a stone dropped 90 m will drift east by 9 mm. Run the test and that's what happens, therefore (and this is the sketchy part) the Earth rotates. That's the same fallacy as concluding a patient has the flu simply because she has a fever. It's the fallacy of affirming the consequent. A fever is some indication of the flu, but it's certainly not proof. Eastward deflection of a falling stone is some indication the Earth rotates, but it's not proof.

More mine shafts were called into the service of this updated tower argument. Reliability was uneven. Most notable was in 1831 when Ferdinand Reich dropped 106 metal balls down a 160-m wooden shaft specially constructed in a mine in Freiberg, Saxony. The average deviation east was 28.4 mm, compared to the theoretically predicted 27.4 mm. In an experiment in a Cornish mine, the drop was a breathtaking quarter mile (400 m), but there were dubious means of determining the true vertical over that long distance. Indoor measurements were shorter but better controlled. A 1902 experiment at Harvard University dropping 948 balls down 23 m recorded an average eastward drift of 0.15 ± 0.005 cm. The predicted value was 0.18 cm, but this was assuming there was no resistance from air.

Terrestrial evidence for the rotation of the Earth was slow and fitful. Rotation is hard to understand, even with a reliable mathematical model to work with. Both the equatorial bulge and the eastward deviation of a falling ball are hard to measure. But by 1800, around the time of the Schlebusch mine experiment and the Svanberg geodetic measurements, there seemed to be good evidence that the Earth rotates, corroboration of the belief in the heliocentric Copernican cosmology. This was 250 years after Copernicus published *On the Revolutions of the Heavenly Spheres*, and almost 200 years after Galileo pleaded his case before the Inquisition.

CHAPTER 11
Foucault's Pendulum

Vous êtes invités à venir voir tourner la Terre dans la salle méridienne de 'Observatoire de Paris.—You are invited to come see the Earth turn, in the meridian room of the Paris Observatory.
—**Léon Foucault**

The Eiffel Tower is inscribed with the names of 72 influential mathematicians, scientists, and engineers, all of them French, chosen by Gustav Eiffel to represent his countrymen's technical contributions that made the tower possible. One of them is Pierre Laplace. Isaac Newton and Gottfried Leibniz may have invented the calculus as a way of describing the mechanics of the natural world, but no one did more than Laplace in applying the math to the details of the phenomena. His five-volume *Mécanique Céleste* is the paradigm of applied mathematics—meticulous, detailed, and comprehensive. It's in the fourth volume that he lays out the calculation of eastward drift of an object in free-fall, his analysis of the tower argument. It starts with this assessment of the situation.

> Although the rotation of the earth is now established, with all the certainty which comports with the state of the physical sciences,

yet a direct proof of this phenomenon must be interesting to mathematicians and astronomers.

This is sometimes interpreted as Laplace still looking for the direct proof, *une prevue directe*, of the rotation of the Earth. Volume four of *Celestial Mechanics* was published in 1805, after the dropped-ball data from the Schlebusch mine, but before the more conclusive results from Freiberg. It could be that Laplace counted the ball-dropping experiments that had already been done as, in fact, the direct proof. Then, just for the sake of bringing mathematicians and astronomers up to speed, he would show them how it's done, that is, show them the math. After 17 pages of elegant calculation, knit together by just a few words of explanation, he arrived at the formula for predicting the magnitude of eastward deviation from a vertical fall. Then he added a brief comment on the state of the testing.

> There have been made, in Italy and Germany, several experiments upon the fall of bodies, which agree with the preceding results. But these experiments, which require very great care, ought to be repeated with still greater accuracy.

This suggests that for Laplace, among the most authoritative experts on the Copernican model of the solar system and Newtonian dynamics, the evidence was still not quite there for a direct proof of the Earth's rotation. That was more than 250 years after the publication of *On the Revolutions of the Heavenly Spheres*.

No matter how good the data are and how closely they agree with the calculated prediction, it's hard to believe that Laplace, or any other tough-minded scientist, would count the measurement of the eastward drift of a falling stone as a direct proof that the Earth rotates. It may be good evidence of rotation, but filtered through the interpretive influence of Newtonian dynamics and the logic of affirming the consequent, it is circumstantial evidence. Laplace would have known of the measurements of equatorial bulge as well, and those data were precise and consistent. But again, this is not a direct proof that the Earth rotates. The *Celestial Mechanics* included some other measurable effects of rotation, and some of those would be the inspiration for future engineers and scientists with innovative experiments. But observing the effects is not observing the cause, and even with Laplace's immaculate calculations connecting the two, the results are not direct proof. This is the place to remember the warning at the beginning of your science textbook that says it is impossible to prove a scientific theory.

Before describing those innovative experiments, we should catch up on the status of the search for stellar parallax. That would be even more indirect; it would be evidence of the Earth's annual revolution around the Sun as a way to support the Copernican model, and, consequently, the Earth's rotation. As more pieces of evidence for the whole system are put together, the case gets stronger for each individual component.

At the beginning of the nineteenth century, the ongoing search for stellar parallax had become a long series of failed predictions by the heliocentric model of the solar system. Even with telescopes, no parallax had been detected. If this is a characteristic example of scientific testing, then the preface to a science textbook should also warn that a scientific theory cannot be disproven either, as long as we are willing to make excuses for the failure of predictions. In the case of parallax, it was the great distance to the stars that made detection of the phenomenon impossible, so far.

Galileo had contributed not only the telescope to the search for parallax, but also a revised technique. With the loss of the heavenly spheres to hold celestial objects in place came the understanding that the stars are not all the same distance away from the Earth. They will consequently have unequal parallax, the effect being larger for the nearer stars. Galileo suggested looking for parallax in a nearby star by using a much more distant star as the reference. This is effectively treating the distant star as if it is infinitely far away.

The technique was to choose two stars that appear very close together, a visual binary, but are in fact at greatly different distances. The challenge is to determine which stars are close and which are far. Stars differ in their apparent brightness, and, intuitively, the brighter stars are the ones closer to the Earth. Add to this the discovery in 1718 by Edmund Halley that some stars exhibit proper motion, that is, a very slow movement against the background pattern. It turns out that the stars are not fixed in place. Once again, intuition would have detectable proper motion in the closest stars. With this in mind, presumptive nearby stars were sought and matched with much more distant visual neighbors to search for parallax.

By the time of Hooke and Newton, the motions of the Earth were generally accepted to be true. The search for parallax was less about confirming the Copernican model than about measuring the distance to the stars. Hooke set up a special telescope, built into a room in his house and directed at the zenith, specifically to detect parallax. Malfunctions and matters of ill health allowed him to make only four measurements with the device, but he nonetheless declared positive results. The astronomical community was generally unconvinced, despite their hope for a declaration of positive parallax.

Using a similar zenith telescope in 1725, James Bradley and Samuel Molyneus studied the same star that had drawn Hooke's attention. Gamma Draconis is bright, and presumably close. Their telescope was set up to pivot slightly, allowing precise measurement along the north-south orientation. The geometry of the star's position and the path of the Earth in its presumptive orbit around the Sun indicated that the parallax would shift to its furthest point south on December 18. Bradley and Molyneus did detect this motion, but were surprised when it continued south, days after the 18th. It continued south until March, and by then it had shifted a phenomenal 20 seconds of arc. Three months later than expected, the star stopped and headed back north, passing through its December point, until September, when it stopped and started south. This was a surprise.

Bradley figured out what was happening. It wasn't parallax, but it was nonetheless evidence that the Earth is moving. Light travels at a fast but finite speed, and in the optical theory of Isaac Newton it consists of a stream of tiny particles. Think of the light from a star as similar to falling rain, particles streaming in at a finite speed. If you move quickly through the rain, the drops appear to come at you at an angle. If you want to collect the drops in a tube, pivot the tube in the direction of your own motion. The tube will be pointing at the source of the rain. Now apply this to light from a star, and, if the Earth is moving, the telescope will have to be tipped forward to catch the particles and to point at the source, the star. Light is much faster than falling rain, so the angle of deflection is much smaller. The phenomenon is called aberration, in this case, stellar aberration. Bradley claimed to have been struck by the idea while sailing and noting that the weathervane on his boat pointed in a direction that depended on both the flow of the wind and the motion of the boat.

Stellar aberration is evidence of the Earth's annual revolution around the Sun. It is 3 months out of phase with parallax since it is caused by the Earth's changing velocity relative to the star rather than the changing position. Aberration was discovered in 1725; parallax had yet to be detected. Since the discovery of stellar aberration came as a surprise, it avoided much of the concern about simply finding what you're looking for and believing the results of an experiment because they confirm your favorite hypothesis. It is sometimes cited as the first clear evidence that the Earth revolves around the Sun.

The initial explanation of aberration, citing an incoming stream of light particles, made a lot of sense, but it was based on an inconsistent theory of light. The particles were described as traveling through empty space, requiring no medium of propagation. Their speed is a universal constant, the speed of light, independent of the source of the light, in this case the star. This is the inconsistency, since the speed of a particle is the composite of its inherent speed plus that of its source. Waves can be generated with a speed independent of the generator, but in the eighteenth century, a wave was thought to require a medium of propagation, as a water wave is nothing without the water. But if light is described as a wave pattern traveling through an invisible, universally pervasive medium—the so-called luminiferous ether—there can be no effect of aberration. Some *ad hoc* remedies were proposed to align a theory of optics with the phenomenon of stellar aberration, most having to do with turbulence or dragging effects in the ether, but none quite worked. Arguably, stellar aberration was not understood until the special theory of relativity provided a consistent account of the absolute speed of light that propagates as a wave in the absence of ether. That was in 1905. For almost two centuries, stellar aberration was accepted as evidence that the Earth moves, without clearly understanding how aberration works. It's worth noting that Bradley calculated the speed of the Earth in terms of the tangent of the aberration angle. The actual formula is in terms of the sine of the angle. It worked for Bradley because with very small angles the tangent and sine are nearly equal.

Meanwhile, the search for stellar parallax continued. With the idea that stars might have significantly different intrinsic brightness, the determination of which stars are near to the Earth depended less on their apparent brightness and more on proper motion. Friedrich Bessel chose 61 Cygni, the sixty-first brightest star in the constellation Cygnus, the swan. It's low on the list and not at all bright to look at, but with its relatively swift proper motion, five arc-seconds per year, it was known as the flying star.

Bessel was both a mathematician and astronomer, and already at the age of 25 he was the director of the Königsberg Observatory. He began his observations of 61 Cygni in 1834, but was interrupted and distracted by the appearance of Halley's comet. He got back to work on parallax in 1837. Measuring parallax takes time, as the Earth moves through its orbit around the Sun, but by 1838, Bessel reported a credible measurement, the star shifting by one-third of an arc second in 6 months. Parallax had at last been observed. Just weeks later, an observatory in South Africa measured a full second of arc shift in the star Alpha Centauri, what we now know to be the nearest star to Earth. Hiding in the southern sky, it had escaped earlier detection.

Friedrich Bessel is the name associated with the first detection of the long anticipated stellar parallax. The celebration of his accomplishment was astronomical. John Herschel addressed the Royal Astronomical Society on the event of honoring Bessel

> Gentlemen of the Astronomical Society, I congratulate you and myself that we have lived to see the great and hitherto impassible barrier to our excursions into the sidereal universe; that barrier against which we have chafed so long and so vainly—(æstuantes angusto limite mundi)—almost simultaneously overleaped at three points. It is the greatest and most glorious triumph which practical astronomy has ever witnessed.

The celestial evidence had been productive, at least for showing the Earth's annual revolution around the Sun. Direct proof of rotation was another matter, and Laplace's *Celestial Mechanics* had more to contribute, and more to say about the terrestrial experiments that could help. In the same fourth volume that did the math for the tower argument, Laplace worked out the effect of a rotating Earth on a projectile shot straight up. Remarkably, he predicted it would fall to the ground a very short distance to the west of the launch. He got this by a thorough analysis of all aspects of the force acting on a projectile. They are dependent on the direction and speed of the projectile motion, and on the rotation of the Earth. There are four distinct components, always perpendicular to the velocity of the moving object. Going up, the projectile experiences a force to the west, the opposite of the eastward force on a falling stone. Going north, the force is to the east. Going east, the direction of the Earth's rotation, the force has two components, south and up. This is complicated, since as an object moves it is continuously changing direction under the influences of components of force, and so the force itself changes continuously. Rotational dynamics present a mathematical challenge, but Laplace's breakdown into these four components made the calculations manageable and revealed more testable consequences of rotation.

The tower argument, whether things go up or down, is just about vertical motion of the projectile. Laplace showed that the horizontal motion of a projectile would also be deflected on a rotating Earth. This had been suggested in 1651 by Giovanni Riccioli who argued that, if the Earth rotated, as he believed it did not, a cannon ball fired toward the north would veer eastward. Detecting no such deviation from the northern trajectory of cannon balls, Riccioli conclude that the Earth does not move. Lacking the means for neither a precise quantitative prediction of the amount of deflection, nor the instrumental wherewithal to measure it, this was not a decisive disproof of the Copernican world system.

By the nineteenth century, both the math and the measuring had improved. In 1835, Gaspard-Gustave Coriolis published a treatise on the forces involved with rotating machine parts, things like waterwheels. This was an application of Laplace's rotational dynamics to a two-dimensional system. Coriolis described what he called a "compound centrifugal force," a combination of the outward force of rotation and the perpendicular component as identified by Laplace. A slider on a rotating disk, for example, would be forced not only out toward the rim—this is the regular Coriolis effect—but also back against the direction of rotation. Its trajectory would be a curve. Later, at the beginning of the twentieth century, this would be known as the Coriolis effect.

Laplace had considered this effect while analyzing the ocean tides. That is, he considered the effect of this compound force on fluids rather than an individual solid object like the slider on a rotating disk. Air is a fluid, and differential heating of the atmosphere causes it to move around. In 1856, William Ferrel applied the idea of a compound centrifugal force, a Coriolis force, to explain the global patterns of prevailing winds, the trade winds. This was, of course, a great help to navigation and meteorology, but it also found evidence of the Earth's rotation in the circulation of the atmosphere. Applying the principles on a smaller scale, to individual storms, offered similar benefits. Solar heating causes air to rise, creating an area of low pressure, into which peripheral air flows horizontally. Laplace's analysis showed that from any direction, the incoming air veers to the right (in the northern hemisphere), thus creating a counterclockwise vortex. The wind always circulates counterclockwise around a low-pressure zone in the northern hemisphere, a Coriolis effect caused by the rotation of the Earth. This is meteorological evidence of the Earth's rotation.

And then there's the water circling the bathtub drain. It's just not true that a bathtub will invariably drain with a counterclockwise vortex in the northern hemisphere and a clockwise swirl in the south. But that's only because the Coriolis effect on such a small scale is miniscule and will be overwhelmed by the most minute turbulence or external influences. If you do it carefully, though, very carefully, the correlation is true. An experiment presented by Ascher Shapiro in 1962 in the prestigious journal *Nature*, reported consistent results following meticulous control. He used 300 gallons (1,100 L) of water, allowed to rest for 24 hours, and a clever mechanism to open the drain without disturbing the water. "When all the precautions described were taken, the vortex was invariably in the counterclockwise direction." After 12 minutes, the water was spinning

30,000 times faster than the Earth's leisurely single revolution per day. Thus did the bathtub act as an amplifier, enhancing the phenomenon of the Earth's rotation to an easily observable magnitude.

Returning to the argument of Riccioli, that a cannon fired north should be seen to curve to the east, bigger guns with longer range proved this to be true. Consider the so-called Paris Gun, a German weapon used in World War I to fire explosive shells into Paris. It was set up 120 km from the target, northeast of the city. Precise aim was never the goal at that range, and the gun was more to cause fear than any specific damage, but at such a distance, the shells curved hundreds of meters to the west while on their southwesterly trip into Paris. Getting anywhere near the chosen target required compensation for the Coriolis effect.

One of Coriolis' teachers, Siméon Denis Poisson, thought to apply the details of Laplace's rotational dynamics to the behavior of a pendulum. He concluded that a simple swinging pendulum would deviate slightly from a straight back-and-forth, due to the rotation of the Earth, but the magnitude of the effect would be far too small to detect. But, in fact, Vincenzo Viviani, a student of Galileo and the man responsible for the story of his teacher dropping two balls, one heavy, the other lighter, from the leaning tower of Pisa, reported an annoying drift of a pendulum he was using in experiments. Always the same direction, clockwise, the pendulum would slowly turn as it continued swinging back and forth. Viviani took this to be a nuisance, remedied by attaching a second rope to the pendulum to stop the drift. He missed making the connection between this behavior of the pendulum and the rotation of the Earth, despite his having worked with Borelli in trying to determine the eastward drift of a falling stone.

Léon Foucault got the connection in 1851, and he is often known as the man who proved the Earth rotates. Each swing of a pendulum will be deflected a very little bit, as Poisson concluded. The change in a single oscillation may be too small to detect, but the effect is cumulative. The key was to let the pendulum continue swinging for a long time, so the predicted effect is gradually revealed. The real challenge would be engineering, building a pendulum with such exquisite balance and the capacity to oscillate for a very long time with minimal damping. It was no longer a math problem but a technological one, and Foucault was perfectly suited to meet the challenge.

Foucault was born in 1819 in Paris. He was a frail child, with a sharp mind and a preference for solitude. With poor eyesight but good hands, he was directed into an education in medicine, hopeful of becoming a surgeon. But he fainted at his first encounter with blood, and a new career path was indicated. He developed an interest in optics and worked with Louis Daguerre on the early development of photography. Teaming up with a friend from college, Hippolyte Fizeau, he designed and built a device to measure the speed of light, an important physical parameter that could be used to test between the competing theories of the nature of light. The Newtonian theory described light as a stream of particles. An alternative, as we encountered while looking for an explanation of stellar aberration, is that light propagates as a wave, more like sound than a rain storm. Foucault cleverly put a mirror on a spinning wheel, powered by a small steam engine, to deflect a beam of light by an angle that depended on its speed. In 1850, he and Fizeau demonstrated that the light travels slower in water than in air, as predicted by the wave theory and contrary to Newton. Foucault could be known as the man, or at least one of them, who proved that light is a wave.

Foucault had the good fortune to discover the intersection of his talents and interests, and to make a life of it. He was to make things, tools to test the fundamentals of scientific theories, and he would be known for what he built rather than what he wrote or theorized. It is also good fortune that among his engineering interests was the pendulum clock.

Discovery comes not just to the observant but to the well-prepared mind. Foucault noticed that a rod of metal spinning in a lathe will vibrate if you tap it, and the plane of vibration stays fixed, even though the rod is rotating. This gave him the idea of the plane of oscillation of a pendulum remaining fixed as its holder, the Earth, rotates. The Earth's rotation is slow, so the effect is small, but it's a little bit with each swing and slowly the pendulum should move, veering to the west, clockwise. When the effect is so small, the apparatus must be isolated from noise and perturbation—recall the bathtub vortex. The pendulum should be long, heavy, and free to swing without any external forces. Foucault started in his basement in Paris, with a 5 kg ball hung on a 2 m steel wire. The wire broke after just a few oscillations. Metal fatigue at the anchor was the cause, since bending back and forth quickly weakened the steel. The technical challenge was finding a way to hold the wire without it wearing out.

Foucault worked it out, and noted the first success in his journal, "Wednesday, January 8, 2 a.m.: the pendulum turned in the direction of the diurnal motion of the celestial sphere." It was easy to convince the scientific authorities of the importance and legitimacy of the demonstration, and Foucault was asked to set up a larger model to present to the public. They chose the meridian room of the Paris Observatory. The larger space could accommodate a longer pendulum, and Foucault prepared an 11-m cable. When the demonstration was ready, he distributed printed invitations to the scientists of Paris. It was a promise of the direct observation that Laplace had been looking for. "You are invited to come see the Earth turn, in the meridian room of the Paris Observatory."

It was a great success. The rotation of the Earth, long known to be true, was now plainly visible, and from the surface of the Earth itself. The public, not just the community of scientists, should enjoy the view as well. So, an even grander display, an even longer pendulum, was hung in the Paris Pantheon, a space that represented the sometimes tense relationship between the religious and the rational. This pendulum was 67 m long, supporting a 17 cm brass ball that was filled with lead. It weighed 28 kg. Such a long pendulum has a correspondingly long period of oscillation, giving the Earth some time to rotate between each return of the ball. With a stylus on the end, the ball struck a mark in sand on each swing, successive marks visibly separated by a little more than 2 mm.

Concerned about the possibility of the steel wire breaking and the heavy ball injuring a spectator or damaging the tile floor, Foucault had installed a fence to keep the crowds at a distance and a cushioning layer of dirt, 20 cm thick. This was a good idea. The demonstration drew large crowds to see the Earth rotate and it continued to swing and veer clockwise for 2 months before, as feared, the wire broke. It was not repaired, and the Pantheon was without a Foucault pendulum until 1902. The latest version now on the display was installed in 1955; it continues to swing today, periodically restarted as the resistance of air dampens the oscillations.

Foucault pendulums quickly appeared in other public institutions of science around the world. It's easy to see for yourself, if you spend a little time, that the pendulum is slowly rotating to the west. And on the surface, the explanation makes sense, that the pendulum is just swinging back and forth while the Earth rotates underneath it. But beyond that, on second thought, it's a little confusing. If you wait around long enough, longer than even the most interested public visitors, it turns out that the pendulum does not return to where it started in 24 hours. The time for the pendulum to rotate fully around depends on the latitude where it swings. In Paris, for example, at 49° north latitude, it takes almost 32 hours for the pendulum to come fully around.

The visual connection between the drifting swing of the pendulum and the rotation of the Earth is easy; the math is complicated. It's easy to believe that Foucault's pendulum shows the Earth rotate; it's difficult to understand the details. The pendulum does not swing back and forth in a plane that stays fixed in space. It is not free to oscillate like the rod in the lathe. The pendulum is constrained to follow its anchor at the top, and that is being dragged along with the Earth's rotation. Each swing of the pendulum gets a small Coriolis curve, to the west as it swings south, to the east as it swings north. This pulls the ball clockwise on each swing, gradually moving it clockwise. If the pendulum was at the north pole, the plane of the pendulum swing would stay aligned with the stars, and the clockwise circuit would take one sidereal day, just under 24 hours. At the equator, there is no veering of the pendulum at all. It swings back and forth in the same plane relative to the Earth; it's just a boring pendulum.

Between the 24-hour period at the pole and the infinite period at the equator, there must be a smooth transition, a dependence of period on latitude such that the period increases continuously as the latitude decreases. We know the endpoints of the function, 24 hours at latitude 90° and infinite at latitude 0°; what's the function itself? It turns out to be remarkably simple, although the mathematical derivation is devilishly difficult. Foucault had it at the start. The time for the pendulum to come completely around is one sidereal day (23 hours and 56 minutes) divided by the sine of the angle of latitude. At the pole, the angle is 90° and the sine is 1. At the equator, the angle is 0° and the sine is 0. In Paris, the angle is 49° and the sine is 0.755, so the period is 31 hours and 45 minutes, as it is measured.

It was not clear how Foucault derived the sine formula, and there was a quick scramble to provide the theoretical foundation for the behavior of the pendulum. Significant disagreements over the details of the reasoning were able to produce the same answer. Some treated the motion as a result of forces and used a dynamic approach; others ignored the forces and applied principles of geometry. The clear derivation and full theoretical understanding weren't available until 1879, in a doctoral dissertation by Heike Kamerlingh Onnes. You might say that Foucault was the man who showed that the Earth rotates, but Onnes was the man who proved it.

Foucault's pendulum was, and still is, a great spectacle and a convincing demonstration of the Earth's rotation. But it's a bit confusing, since it does not stay fixed in space and it does not return to its starting point at the same time each day. More straightforward would be a device that spins on its own axis rather than oscillates on a plane. Allow the axis free movement and it will in fact remain aligned in space no matter how its support is moved. Such a spinning sphere had been invented in 1817, and Laplace was known to use it as a lecture demonstration. Foucault adapted it in 1852 (he was not one to sit around) as a spinning torus with the center axis in a frame with two gimbals to allow the freedom of orientation in any direction. He called it a gyroscope. With a hand-crank and gears, the gyroscope could be brought up to speed at an impressive 200 revolutions per second. It would continue to spin for up to 10 minutes. The Earth doesn't rotate very much in 10 minutes, and you won't notice a gyroscope veer to the west if you use just a toy, but Foucault had his fixed up with a delicate needle-mirror-microscope system to detect the slightest reorientation. Consistently, the gyroscope veered west, with a timing to bring it around in one sidereal day, at any latitude.

There are now hundreds of Foucault pendulums on public display around the globe, even one at the Stag's Leap Wine Cellars in Napa, California. There are untold numbers of gyroscopes deployed as direction-preserving navigation devices. Foucault's legacy is public and profound. The Foucault pendulum is the most common first response to the question, How do we know the Earth rotates? You are invited to see for yourself, at a science museum or public observatory, even, if you are in northern California, with a glass of wine.

Foucault's name is among the 72 on the Eiffel Tower. It's up there with Laplace. So is Coriolis.

The Postmodern Perspective

CHAPTER 12
Mach's Principle

These two propositions, "the earth turns round" and, "it is more convenient to suppose that the earth turns round," have one and the same meaning.
—Henri Poincaré

There are a few famous scientists you would expect to show up in a discussion of the rotation of the Earth. Copernicus is certainly at the top of the list. Galileo is there. Foucault, *nom du pendule*, is likely, at least if you've been through the lobby of a well-funded science museum. And, anticipating the reckoning of relativity, Einstein is on the way. But Ernst Mach may be a surprise. His name is familiar, but in a different context. It's about the speed of sound, as in Mach-one, not about testing the Copernican model of the solar system.

There is no sonic boom announcing the speed of the Earth. It doesn't even make sense to say it's supersonic, since any medium that could carry a sound either moves with the Earth in its daily rotation, or is all together absent from its yearly revolution. Mach's role in this debate has nothing to do with sound or sonic speeds; it's about his alternative interpretation of the otherwise convincing evidence already on the books that the Earth rotates. Foucault's pendulum,

Mach will argue, does not at all show the rotation of the Earth. The equatorial bulge is equally uninformative. And even the concave surface of the water in Newton's bucket is not evidence that it's spinning, at least not in the absolute sense. This is reminiscent of Galileo's warning that nothing we can observe on the Earth itself can reveal our planet's motion. It's the principle of relativity again, and Mach will be more faithful to it than was Galileo.

Mach's analysis starts with a strict understanding of the role of scientific observation, and a cautious guideline on the limits of scientific conclusions. As methodological motivation this will both challenge the basic claim that the Earth rotates and get us to the theory of relativity.

Ernst Mach lived from 1838 to 1916, a lifetime that included some of the most important developments of modern physics, more or less when the textbooks now used by physics graduate students were written. The theory of the atomic composition of matter, for example, had just been introduced and was being refined and merged with the larger understanding of the natural world. Mach refused to endorse the atomic theory, and he never believed in the reality of atoms. In this way he was in a minority among nineteenth-century physicists, but he was not shy in his dissent. He is said to have shouted from the audience during a lecture by his colleague Ludwig Boltzmann, on the correlation between random motions of atoms and temperature, "I do not believe that atoms exist." When asked why not, Mach replied rhetorically, "Have you seen one?" (*Haben Sie einen gesehen?*) This outburst may or may not have happened, but the sentiment is true to Mach's core ideas on scientific method. It is consistent with both his character and his commitment to direct observation. Similar to the story about Galileo's, "And yet it moves," true or not, it's telling.

Many working scientists endorse this sort of uncompromising empiricism when they talk about method and clarify the credentials of the scientific process. Stick to the facts, the things that have been seen. But the tough talk softens in practice, as it must. It's not that scientific conclusions are *restricted* to what can be observed, as Mach required; rather, they must be in some accountable way *based on* what can be observed. There's a lot of latitude in the concept of based-on, and by now the majority opinion in physics goes well beyond belief in the existence of atoms. From dark matter to gravity waves, Mach would have a lot to shout about at a conference on cosmology.

A lot of physics happened in Mach's lifetime. He was born in 1838, the year stellar parallax was first measured. One year before Mach's death in 1916, Einstein presented the general theory of relativity, the new account of gravity and the replacement of Newtonian gravitation. Between those years, modern physics was made, and much of it involved things that could not be seem, at least not directly. As it always had, science continued to reveal the ways in which the reality of nature is different from the way things appear.

The modern atomic theory that Mach doubted was introduced in 1805 by John Dalton. By noting recurring patterns in the proportions of constituents of compounds like metal oxides, Dalton reasoned that the elements exist in discrete pieces, atoms. Every atom of a particular element like oxygen or hydrogen is identical and indestructible and characterized by its mass. Dalton began a list of elements in order of their atomic masses. Dmitri Mendeleev noticed a pattern in the list, and in 1869 produced the periodic table of elements. This first periodic table included sixty elements; there are now over one hundred. Remarkably, Mendeleev left some gaps, unfilled positions in the table where no element had been observed but for which he could predict the properties. Subsequent discoveries filled in the gaps, as predicted.

At the same time, modern thermodynamics developed in ways that both facilitated an industrial revolution and incorporated the theory of atoms. Great technological progress had been made with the theory that heat was an actual substance that flowed from a place of high temperature to lower, just as water flows downhill. It was called caloric fluid. By this account, a hot potato contains more caloric fluid than a cold one, and an oven is a caloric-fluid-delivery appliance. It made sense, and the first law of thermodynamics was written in terms of the caloric theory of heat. But new experiments at the end of the eighteenth century and the beginning of the nineteenth suggested that heat is not a separate substance but a property of the material already there. Hot and cold potatoes have exactly the same ingredients; it's just that the atoms and molecules in the hot one have a more energetic random motion than in the cold. This is the kinetic theory of heat. It is equally consistent with the first law of thermodynamics, now written in terms of conserving energy rather than caloric fluid.

It's one thing to theorize in general about the kinetic energy of atoms and the measurable phenomenon of temperature; it's another thing to make the correlation precise and quantitative. In 1859, James Clerk Maxwell did the math, showing that a variety of speeds in the atoms of a gas could account for various thermal properties. Since the different speeds are distributed over a spectrum of values, thermodynamics became linked to what is now called statistical mechanics. The kinetic theory took over, and the second law of thermodynamics took shape in terms of random motion of the unseen particles.

This sort of unification and coherence, bringing very different phenomena under a single explanatory theory, is a mark of progress in science. Some would argue that it is also an indication that the unifying theory is true. It is surely an improvement in the economy of description, and it was happening in dramatic fashion during Mach's career.

Optics, the study of light, experienced a conceptual reformation similar to the progress in thermodynamics. It was a house divided between two theories on the fundamental nature of light. As with heat, one theory attributed the phenomenon of light to the presence of a specific substance, a stream of particles rather than a fluid. The alternative idea was that light is a property of material already there, a pattern of waves moving through an ethereal space.

The wave theory seemed the winner when, in 1801, Thomas Young passed light through two narrow, closely spaced openings and projected not just two bright spots, as a stream of particles would produce, but a spread-out pattern of recurring bright and dark spots. This demonstrated interference, a behavior distinctive to waves. It showed the peaks in the waves from one opening meeting the peaks from the other—creating the bright spots, or meeting the troughs from the other—creating the dark. In retrospect, many textbooks now describe this as the definitive evidence, but good scientists know that no single experiment either proves or disproves a hypothesis, and the stream-of-particles characterization of light hung around into the twentieth century.

More difficulties for the light-particle theory came in 1850. The wave theory predicted that light would move slower through water than through air; the particle theory said it would go faster in water. Recall that it was Léon Foucault and an associate Hippolyte Fizeau who devised ways to measure the speed of light with unprecedented precision. The new techniques of measurement clearly showed light went slower in water than in air.

Light as a wave pattern seems to require a medium. Sound is a wave, but it won't travel in a vacuum. It moves through material like air, and Mach eventually made photographs of sound waves, including the shock wave created when the source of the sound is moving at the speed of the wave it produces. But there was no visible record of a medium for light waves, only the illumination of the source and the eventual reception of the light, nothing in between, and nothing with peaks and troughs. So what is it that waves, eventually showing up as light transmitted from source to our eyes?

Unexpectedly, the answer came from studies of electricity. This is an heady case of unification and coherence, first between electricity and magnetism, and then including light.

In a series of empirical discoveries and theoretical insight, the two distinct forces of electricity and magnetism were revealed to have a common source, electrically charged objects. The two forces differ only in the context and circumstances of the charged object, or, as physicists abbreviate it, the charge. In the simplest circumstance, an electric charge creates a force on any other charge; like charges repel, unlike charges attract. A *moving* charge (and this is where the story becomes relevant to the question of whether the Earth is moving) creates an additional force. This dynamic electricity is magnetism.

The discoveries of electrodynamics—and that's the term for the topic studied by today's physics students—came in quick succession. In 1820, Hans Ørsted found that a magnet deflected when electric current passed through a nearby wire. The following year, André-Marie Ampère demonstrated a force between two wires when both conducted an electric current. In 1831, Michael Faraday discovered electromagnetic induction; a changing magnetic field creates an electric current. This is how an electric generator works, rotating a loop of wire in the field of a strong magnet.

These are just a few highlights in the evidence of reciprocity between electricity and magnetism. Again it was Maxwell who attended to the details and put the pieces of the puzzle together. In 1864, he wrote a quartet of equations tying together the dynamic behavior of electric and magnetic fields. The result was more than the sum of the parts, in that oscillations in one field produce oscillations in the other, in turn producing a self-sustaining wave that propagates out from the source. This is electromagnetic radiation, electromagnetic waves. In 1873, Maxwell demonstrated that light is one form of electromagnetic radiation.

This is a tidy theoretical package covering a diverse collection of phenomena, from electricity and magnetism to light, and it will go on to include radio signals and radioactivity. But there are a lot of things about electrodynamics that defy observation. Have you ever seen an electric field? Have you ever seen the wave pattern that propagates as electromagnetic radiation? At the time, it was assumed that these waves moved through a pervasive medium termed the luminiferous ether, but its properties were enigmatic, since only a solid medium can support transverse, side-to-side waves like the electromagnetic waves, and only a very dense solid can facilitate the very fast speed of light. How could a pervasive dense solid be otherwise undetectable? And how could other solid objects, things like planets and stones, move uninhibited through the solid ether?

A dose of Viennese skepticism, with its down-to-earth insistence on evidence, is probably valuable at a time of fundamental changes in a science. The excitement of new ideas may be charming, and it takes a curmudgeon like Mach to challenge the novelties like atoms and electromagnetic fields. The purpose of science, as Mach saw things, is to provide an economical and pragmatic description of the natural world, free of superstition and any idle, metaphysical concepts like atoms or ether or, let's not forget, crystalline celestial spheres.

But physics already had a tradition of dealing in things unobservable, some on the edge of occult. Newtonian mechanics included an essential entity that not only hadn't been observed but, in principle, never could be observed. It's what Newton called absolute space, "in its own nature, without regard to anything external, remains always similar and immovable." This is the unobservable container of everything in the universe, and the reference for absolute motion. It's what holds still while the bucket rotates, and through which the Earth moves. It is a requirement for determining that the Earth rotates in more than the relative sense, but, Mach would have to ask, have you seen it?

It's not clear how Mach came by his strict rejection of things unseen. He was born in Moravia, then under control of the Austrian Empire, now a part of the Czech Republic. His was an academic family, and he was initially home-schooled. The natural academic momentum took him to the University of Vienna where he earned a doctorate in physics. His doctoral thesis had the title, *Uber elektriche Landungen und Induktion,* On electric charge and induction. Unlike most scientists at universities today, his thesis topic did not set the research boundaries of his career, and his principle contributions to physics were not about electrical induction but about sound and light, and the physiology of perception. He taught physics to medical students while at the University of Vienna, then moved to Graz and on to Prague where he was a professor of experimental physics. He returned to Vienna in 1895, appointed as the Chair of the History and Philosophy of Inductive Science. The word "induction" is playing multiple roles in the account of Mach's professional life. In the title of his thesis it refers to the effective link between the motion of electric charge and a measurable force. It's about evidence of something moving. In the title of his last academic appointment it refers to the logic of drawing conclusions from evidence, induction in contrast to deduction. Induction, in this logical sense is inherently uncertain.

In 1867, Mach turned his attention to mechanics, the most fundamental, and many would say least exciting branch of physics. His interest was in determining the properties of bodies in motion strictly in terms of exactly that, bodies in motion relative to other bodies, without reference to abstract or unobservable quantities. He figured a way to define the mass of an object, which, strictly speaking you can't see, in terms of its relative acceleration, which strictly speaking you can see and measure. It's worth noting that the idea of mass still puzzles scientists; the latest attempt to explain the property relies on the so-called Higgs boson. Mach submitted his work on mass and acceleration to the very prestigious journal, *Annalen der*

Physik. The paper was rejected. Sixteen years later, Mach published a more comprehensive book-length treatise on mechanics, with the sober title, *The Science of Mechanics.* What you see is what you get. This was a historical account of the science, and an explicit challenge to Newton. It had consequences for interpreting the evidence of the rotation of the Earth.

Newton's laws of motion and gravitation had two essential roles in making sense of the evidence of rotation. The force of gravity that holds the solar system together doesn't allow for the old world system, only the new. The compromise Tychonic system, with planets orbiting the Sun as the Sun orbits the Earth, is dynamically impossible, given the gravitational force of attraction that holds the Newtonian system together. The old and new cosmologies may be *geometrically* identical, the same on paper, but they are not mechanically identical. And, as Galileo insisted, it's not the world on paper that counts. Thus does Newtonian mechanics rule in favor of the new world system and, indirectly, for the rotation of the Earth.

The other help from Newton was more direct; the equatorial bulge, like water in a spinning bucket, demonstrates rotation, absolute rotation.

There was more or less universal confidence in Newtonian mechanics in the nineteenth century. Not only had it made tidy sense of things in motion, both celestial and terrestrial, it continued to make predictions of astronomical events with uncanny precision and accuracy. By its predictive authority, a new planet, Neptune, was discovered in 1846. Under the influence of this victory, it was assumed that, in 1859, when the predictions for the orbit of Mercury began to miss what was observed, there must be yet another unseen planet in the mix. The search was on for Vulcan, but not for any theoretical changes to Newton. Despite some bogus sightings, Vulcan was never found, and as far as we know, doesn't exist. We now realize that the fault was not in our planets but in our theory. But at the time, for Mach to challenge Newtonian mechanics was to swim against a strong tide of consensus. We've seen that happen before in the history of science, with revolutionary results that are now celebrated.

Mach was certainly not the first to take issue with the fundamentals of Newtonian mechanics, but as a catalyst for the subsequent theories of relativity, he is the most effective for understanding the relativity of rotation. Two things bothered Mach. First, the idea of gravitation invoked an unseen force acting instantaneously at great distance between two objects. The pejorative term for this is action at a distance. It is exactly what Galileo found objectionable about Kepler's explanation of the tides; it invoked an invisible interaction between the Moon and the oceans. Galileo called it occult. Mach, commenting on both the mystical nature of the theory of gravity and the tendency of scientists to become complacent over time about theoretical problems, put it this way.

> The Newtonian theory of gravitation, on its appearance, disturbed almost all investigators of nature because it was founded on an uncommon unintelligibility. People tried to reduce gravitation to pressure and impact. At the present day gravitation no longer disturbs anybody; it has become *common* unintelligibility.

The theory works, Mach agreed, but that's all you can say. That doesn't mean it's true. Caloric theory worked; it was eventually rejected. If Newtonian theory is rejected as false, we will have to revisit its interpretation of rotation evidence.

The other profound problem in Newton's science is the role of absolute space. For Newton it's a real thing. He is explicit in his more philosophical writing about the mechanics of nature, but the concept of absolute space is only implicit in the mechanics itself. This allows working scientists to ignore it and get on with the work of predicting and measuring. Shut up and calculate, as the saying goes in graduate school.

Mach, as we've seen in the outburst over atoms, was not one to shut up. He pointed out the essential role of absolute space in distinguishing reference frames in which the laws of mechanics apply, the so-called inertial reference frames, from those in which they don't. Noninertial reference frames are those that accelerate in absolute space. Rotation is a form of acceleration. In these frames, fictitious forces show up, forces with no apparent source, centrifugal force, for example, with no cause other than rotation with respect to … absolute space.

Newton, and philosophical Newtonians, admitted that while absolute space cannot be directly observed, its undeniable effects can be. There is a real meaning to the concept of absolute motion. It is real but undetectable in an object with constant speed and direction. It is real and detectable in an object with variable speed or changing direction. It is real and detectable in the spinning bucket. The shape of the surface of the water, they argued, unequivocally indicates rotation. It's not rotation with respect to the bucket, or the building, or the observer, but with respect to space, absolute space. Continue the thought-experiment and imagine the bucket in empty space. (Note our comfort with the phrase "empty space." It reveals a disposition to adopt the Newtonian metaphysics that there is such a thing as space itself.) With nothing around, the water would nonetheless be flat when it does not rotate and concave when it does.

It's easy to imagine Mach's objection. Not only is the evidence circumstantial, it calls for speculation. In science, as in a court room, we need to stick to the facts. What, in fact, do we see? We observe that relative motion between the water and the sides of the bucket has no effect on the shape of the water's surface. Before the experiment begins there is no relative motion between these two and the surface is flat. When the bucket starts to spin, but before the water catches up, there is relative motion, and the surface is flat. Once the water catches up there is again no relative motion, but this time the surface is concave. The shape of the water surface reveals nothing about its rotation relative to the bucket.

Expanding the perspective, we observe that relative rotation between the water and ourselves, the walls of the room, the Earth and the rest of the cosmos *does* have an effect. But, of course, it's not about us, and our place in that list of references is incidental. We could follow around with the spinning water by putting ourselves on a carousel with the bucket at the center and show that *our* relative rotation amounts to nothing. Conceivably, we could put together an experiment in which the whole building spins with the bucket. The water would be concave when the gigantic apparatus rotated, even though there is no rotation relative to the walls. This is speculation, but it seems very reasonable, and no one disputes the projected outcome of the experiment.

Mach pushed the thought-experiment further. Rotate the water with an enormous amount of mass going around with it, more than just ourselves and the building. Since this is so far beyond what has been or could be done or observed, he concludes

> No one is competent to say how the experiment would turn out if the sides of the vessel increased in thickness and mass till they were ultimately several leagues thick.

It's possible that with no rotation between the water and the enormous mass of the vessel there will be no centrifugal force and no curve in the surface. All we can observe in the actual experiment is relative rotation between water and real objects. There is only speculation about what happens when the relative rotation is eliminated.

The thought-experiment only serves to introduce some doubt about Newton's interpretation of the spinning bucket. Mach then proposed his own conclusion, one that followed the rule of avoiding unnecessary unseen entities in scientific theories. The actual experiment shows the water is concave when it rotates with respect to the building, the Earth, and the stars—all real things with real mass. Strictly speaking, that is, according to Mach, scientifically speaking, all that can be reported is this relative rotation, relative to the "fixed stars." All we *see* is relative motion, so all there is relative motion.

This is not simply changing the wording in describing the bucket experiment; it's changing the physics. There is a new force being proposed, a real force, pulling the edges of the water out and up the sides of the bucket when it rotates relative to the cosmic mass. We know, and Mach acknowledged, that things with mass create a force on each other. This is gravity, and it requires real objects. Mach's interpretation of the rotating bucket proposed an additional aspect of gravity, one that shows up only when there is relative acceleration between real objects. It is a gravitational interaction between the water and distant stars that causes the centrifugal force, a real force brought about by the relative motion between the two things. There is no reference to absolute space in this explanation of the concave water surface, only to real material things and, importantly, a relative rotation between them.

All this means that it is just a choice of words and point of reference to say that the water in the bucket is rotating. There are two equally legitimate descriptions of what is observed. When the surface of the water is concave, the water is rotating (with respect to the material constituents of the cosmos, the fixed stars), or, another way to say the same thing, the stars are orbiting the water. The second account seems crazy. It's as if by giving the bucket a whirl, twisting up the rope on which it hangs and letting go, you in fact put the stars in motion, all of them. You can do this at home, with neither bucket nor water. Just stand up and turn around. This makes everything in the universe orbit your (relatively) stationary body. It would seem to require superhuman power.

The apparent implausibility comes from a misleading, if implicit, slide from the fact that there are two distinct descriptions of the situation to thinking there are two distinct situations, one in which you spin and another in which the universe orbits (and so needing an extraordinary driving force to get started). But there is only one situation. There is relative rotation between observed objects, and we are free to choose either as the point of reference. We are used to this with straight-line motion, that is, inertial motion. On a train we can say that we are moving while the Earth-bound scenery stands still, or that we are stationary as the trees and mountains move, because there is only relative motion. There are two descriptions (more actually, since any object, not just train or terrain, can serve as a fixed point of reference) of the one situation. Changing from one to the other doesn't raise any question or concern about a force needed to move mountains.

Now apply this analysis to the evidence that the Earth rotates. The equatorial bulge predicted by Newton and detected in 1736 is a planetary spinning-bucket frozen solid. It is evidence of rotation, but by Machian standards only relative rotation. Say that the Earth rotates while the stars stand still, or say that the Earth is motionless as the stars orbit in 24 hours, and you've said the same thing. The only difference is the choice of reference. Without invoking the mystical, unscientific notion of absolute space, there is no absolute rotation. Since there is no evidence of absolute space, there can be no evidence of absolute motion. Thus spoke Mach.

The Foucault pendulum is more complicated. It's not just a greater challenge to explain on Mach's terms of strictly relative rotation; it's tricky both mathematically and intuitively even assuming absolute space as a reference. That's because there more moving parts, both a rotation of the planet and the periodic swing of the pendulum, and because the action happens at various latitudes, not just the equator. Mach's account of the slow precession of the pendulum is this. The relative motion of the cosmos, the diurnal orbit of the stars, alternatively described as the diurnal rotation of the Earth, creates a force, a gravity-related drag that tugs on each swing of the pendulum in the direction of the cosmic orbit. This is in line with Foucault's first description of the pendulum's precession, "the pendulum turned in the direction of the diurnal motion of the celestial sphere," as if the orbiting cosmos is pulling the pendulum around. Exactly the same force acts on a gyroscope, holding it aligned with the distributed mass in the universe, the stars, and galaxies. The Foucault pendulum differs from the gyroscope in that it is constrained to swing back and forth. It gets a small tug from the stars on each swing, slowly moving it around. Neither the precession of the pendulum nor the alignment of the gyroscope has anything to do with orientation in space. It has everything to do with orientation, and changing orientation, in the cosmic mass. The pendulum does show the Earth rotates, but only relative to the stars.

Mach is not saying that the Earth is stationary. He is not claiming that the old world system of Aristotle, Ptolemy, and Tycho is true while the new Copernican system is false. The point is that the old and new are exactly the same system. In his own words

> The motions of the universe are the same whether we adopt the Ptolemaic or the Copernican mode of view. Both are equally *correct*; only the latter is more simple and more *practical*. The universe is not *twice* given, with the earth at rest and the earth in motion; but only *once*, with it *relative* motions alone determinable.

We are free to describe it with the Earth rotating relative to the stars, or the stars in orbit relative to the Earth.

Since Mach's suggestion is that the evidence does not favor either world system over the other, it is reminiscent of what Galileo made of the vertical fall of a dropped stone, the tower argument. The cases are similar in that what seemed to be decisive evidence turns out to be equivocal. The evidence favors neither of the world systems. Galileo and Mach both did this by using a principle of relativity. It is possible to detect relative motion between two objects, but not possible to determine that one of them is stationary. There is no way to determine the motion of a single object.

There are important differences between the two arguments. Galileo has an implicit restriction to uniform motion, constant speed in a straight line; Mach has no such restriction. And for Mach, it's not just that we can't *detect* the difference between a rotating and stationary Earth; there *is* no difference. But Mach, unlike Galileo, has no persuasive analogy to support his case, no experiment with stones dropped on moving ships. His reasoning is based on the insistence that there is no absolute space that is a reference for motion. And this derives from the methodological principle that what cannot be observed cannot be real. Mach is strict in enforcing a rule that almost all physicists recite at some point in describing their science; keep it as simple as possible and theorize with as few entities as possible. The evidence of equatorial bulge and Foucault pendulum can be explained without absolute space, what Mach regarded as one of the "arbitrary fictions of our imagination."

The evidence can be explained as merely relative rotation, but there was work to be done. The force that pulls the Earth out at the equator and that gently tugs the Foucault pendulum on each swing is an entirely new force introduced by Mach. It can't just be gravity, since it's not there without the relative rotation. Somehow, relative rotation between the Earth and the stars, or between the water in the bucket and the stars, creates an additional force. But we can't leave it at "somehow"; Mach needs details, otherwise this new force will be just another of the arbitrary fictions of our imagination.

The science of gravity we are taught in high school, and what is studied in getting a bachelor's degree in physics, is pretty much what was on the books throughout the nineteenth century. It's the Newtonian theory. There is an attractive force between any two objects that depends only on their masses and the distance that separates them. It does not depend on any relative motion between them. It's analogous to the force of electricity, except it is always attractive and consequently can never be neutralized. A force like this can cause a bulge, of sorts. It is, after all, responsible for the ocean tides. A high tide is a bulge in the water, directed toward the Moon or, because of the decreasing strength on the opposite side of the globe, away from the Moon. This would happen whether or not the Earth and Moon were in relative motion. In the event of no relative motion, that is, if the Moon hovered like a geosynchronous satellite, the high tide would always be in that same spot on the Earth. Tides in fact come and go because the Earth and Moon are in relative motion; the Moon orbits the Earth, or the Earth rotates under the Moon, as you choose.

This can't be what's causing the Earth's equatorial bulge. Unlike the tides, that bulge is evenly distributed around the globe at all times. Furthermore, the mass of the contents of the universe, and hence the gravitational attraction, is more or less evenly distributed. It would pull out along every radial direction of the Earth, not just the equatorial. The poles, and latitudes in between, would bulge just as much. The Earth would be just a tiny bit bigger, but still spherical.

To make Mach's principle of relative rotation square with the phenomena of equatorial bulge and the Foucault pendulum, there has to be an additional component of gravity. Mass causes the gravitational force of attraction we're used to; mass in motion causes something more. Straight, uniform motion between masses won't do it, only accelerated motion, like going around a bend or around in circles. There are, in other words, two manifestations of gravity. There is a static force of attraction between masses, and there is a dynamic force between masses in relative acceleration. Centrifugal force, often dismissed in the textbooks as fictional, is very real in Mach's account. It's a dynamic gravitational force created by relative rotation.

Newton clearly described the phenomenon of static gravity; he wrote the law and put it in precise mathematical form. Mach, in contrast, never filled in the details on his proposal of dynamic gravity. Science is a show-me-the-evidence enterprise. Physics adds to that, show-me-the-math. For Mach's new component of gravity, the evidence is there; it's in the curved surface of rotating water, the equatorial bulge, the Foucault pendulum, and gyroscopes. It's there every time you experience turbulence on an airplane. That feeling in your stomach is a tug from distant stars that are jerking back and forth and creating the dynamic component of gravity. But how, exactly, is this force created? How does it work? What is the precise relation between the masses and distance and acceleration? Show me the math. Mach never did.

The nineteenth-century developments in electrodynamics suggested that it was worth trying. The interaction between electric charges, recall, has two manifestations, the static electric force of electricity and the dynamic electromagnetic force between charges in relative motion. Mach had written his thesis on electromagnetic induction. The analogy, and the goal of a unified, coherent account of nature, suggested a similar structure between electrodynamics and gravitational dynamics. With Mach, though, it remained an unfulfilled project.

Regard for Mach and his ideas on the relativity of rotation was divided. He was either a crank or a profit. Albert Einstein took it as inspiration and motivation, on the way to a detailed theory of what he hoped would be full-on relativity. It was Einstein who introduced the phrase, "Mach's Principle," the requirement that all manner of motion is only relative to real physical objects, with no reference to space itself. Mach was just the beginning, "a catalogue and not a system." For Einstein this was both a criticism and a challenge to figure out the system.

Back to the science class and the two competing statements: The Earth rotates. All motion is relative. At the end of the nineteenth century, more than 400 years after Copernicus published *On the Revolutions of the Heavenly Spheres*, there was some uncertainty on how to reconcile these ideas. Ernst Mach concluded that the second, the relativity of motion, is true without qualification. The first, the rotation of the Earth, is misleading. Sure, the Earth rotates, but only relative to other objects in the universe. But then the declaration is trivial.

CHAPTER 13
Relativity

Rotation is thus relative in Einstein's theory.
—**Willem de Sitter**

... besides observable objects, another thing, which is not perceptible, must be looked upon as real, to enable acceleration or rotation to be looked upon as something real.
—**Albert Einstein**

At the beginning of the twentieth century, there were no scientists on record to advocate for the old world system, the model with the Earth absolutely at rest while the cosmos revolved around in 24 hours. But there were some who opposed the new world system, insofar as it put the Earth in absolute diurnal rotation with the mass of stars held motionless. There still are. They are advocates of Mach's principle who argue that Einstein's theory of relativity is true to Mach and describes rotation as a relative property.

We have seen it frequently before, the idea that only relative motion is admissible in cosmology. Al-Biruni in the tenth century,

"For it is the same whether you take it that the Earth is in motion or the sky. For, in both cases, it does not affect the Astronomical Science." Al-Tusi in the thirteenth century said the same, and Oresme in the fourteenth century, "I suppose that local motion can be perceived only when one body alters its position relative to another." But all of these scientists came to their relativistic conclusions before there was terrestrial evidence for the rotation of the Earth, the centrifugal force of equatorial bulge and the Coriolis effects on projectiles and pendulums. And all of these pre-Copernicans believed that there is absolute rotation in the universe, and the Earth is in fact at rest. Their point was that we just can't perceive that the Earth is stationary, or find any evidence for it, but the fact remains; the Earth does not rotate. The medieval arguments that concluded the Earth is absolutely at rest were grounded in metaphysics, not physics. Mach would say the same for arguments that conclude the Earth is absolutely rotating.

Newton's approach to the question of absolute rotation was to separate movement into two categories, reminiscent of Aristotle's difference between natural and violent motion. Newton distinguished between moving with uniform speed in a straight line, and moving with any manner of acceleration, that is, changing speed or direction. Both are absolute but only the latter is detectable. Both can only be perceived relative to other objects, but there is indirect evidence of accelerated motion—the rotating bucket, for example. Mach argued that the indirect evidence could always be reinterpreted in a way that avoids the mystical invocation of absolute space as the necessary reference for absolute rotation; it is simply evidence of acceleration relative to other objects in the universe. This was a promissory note on which Einstein hoped to make good.

This is the time to take seriously the second part of the lesson from the science class, that motion is relative. It's not just relative as far as you can tell, or all you can observe, but really *is* relative. But if this is true, then such a serious search for evidence of the Earth's rotation has been a waste of time. We've known all along that the Earth rotates relative to the cosmos; the inquisition is about absolute rotation.

A properly named theory of relativity would suggest that all motion is relative. We have to ask whether Einstein's theory of relativity in fact makes this claim. So, what is Einstein's theory of relativity? First, what it is not. The exuberant slogan, "It's all relative!" is not only a false summary of the scientific theory of relativity, it's self-contained nonsense. If we were to distinguish this statement with a title, call it, say, the doctrine of relativity, then, if everything is relative, so is the doctrine of relativity. It severs its own connection to the concepts of true and false. And with all descriptive statements rendered relative and unstable, we lose the ability to reason, communicate, and share our ideas about nature. We lose science.

Happily, this is not at all what Einstein's theory of relativity says or is about. In fact, Einstein objected to the name of the theory; it would have been called the theory of invariance, if he had been able to control the spin. It is principally about things that are not relative. The core idea in the theory of relativity can be traced back to Galileo and the tower argument. He argued that the motion of the Earth, like the motion of a smoothly sailing ship, cannot be detected by experiments on the Earth or the ship itself. That's because the results of all experiments are the same, whether the thing is moving or not. The experiments are done and measurements are made in the reference frame of the Earth or the ship, and the outcome is invariant, the same in a moving reference frame as it is in a stationary frame. And this is simply because the laws of nature are the same in a moving reference as in one that is not moving. This is a principle of invariance; the laws of nature are invariant from reference frame to reference frame. The laws of nature are not relative. This is pretty much what we mean by a law, a regularity with universal application.

This became known as the principle of relativity. It makes some sense. If the motion, or not, of your reference frame cannot be detected by measuring parameters within the frame itself, motion can only be seen relative to other things, other reference frames. Only relative motion is measurable.

Einstein produced two theories of relativity; the difference is rooted in the Newtonian distinction between straight-line uniform speed and motion that changes in speed or direction. The modern terms for these are inertial and noninertial motion, respectively. The work we've done with centrifugal and Coriolis forces suggests that you *can* detect noninertial motion from within the reference frame. It's as if there are different laws in play that give rise to different forces. This will make it difficult to enforce the principle of relativity in a rotating reference frame, like the Earth. That's the goal, under the influence of Mach, but Einstein's first step was to ignore the noninertial complications. Restricting attention to smooth, straight-line motion, it seems easy to apply the principle of relativity. That's just what Galileo did. It was easy until the new laws of electricity and magnetism, the ones that linked electromagnetic waves to light. Then it was confounding, and this was just when Einstein was coming of age.

Albert Einstein was born in Ulm, Germany, in 1879. The family moved to Munich 1 year later, where his father began a business selling electrical equipment. Albert was expected to eventually join the business and was set on the appropriate educational track. The Luitpold high school, now called Albert Einstein Gymnasium, followed a strict and authoritarian protocol of classes, a regimen that Einstein resented and found uninspiring. It didn't help when his family left for business opportunities in Italy, leaving their 15 year old son on his own to finish school. He didn't. He dropped out. This adds Einstein's name to the roster of dignitaries in the science of the Earth's rotation who were abandoned in some way by their father and along the way left school with no degree. Copernicus was raised by an uncle and studied at several universities without graduating. Tycho Brahe was kidnapped and raised by an uncle and toured a variety of universities, earning a degree from none. And Kepler, whose father left the family behind just years after the birth of his son, quit Tübingen University before finishing his studies. Galileo doesn't quite make the list; he dropped out of medical school, but his father never left him. Newton also only half-qualifies, with no father at all but an efficient career and degree from Cambridge University.

The teenage Einstein rejoined his family in Italy and then made his way to Switzerland where he applied to the Swiss Federal Polytechnic. He failed the entrance exam. Realizing that a graduate from a Swiss high school would be automatically admitted, he returned to school and found the Swiss style more to his liking. He not only graduated, he renounced his German citizenship. It would be 5 years before he became a citizen of Switzerland.

Enrolled in the Swiss Federal Polytechnic, Einstein again found the pedagogy uninspiring and often skipped classes. He described the situation in his autobiography,

> ... one had to cram all this stuff into one's head for the examinations, whether one liked it or not. This coercion had such a deterring effect [upon me] that, after I had passed the final examination, I found the consideration of any scientific problems distasteful to me for an entire year. ... It is, in fact, nothing short of a miracle that the modern methods of instruction have not yet entirely strangled the holy curiosity of inquiry; for this delicate little plant, aside from stimulation, stands mainly in need of freedom; without this it necessarily goes to wreck and ruin. It is a very grave mistake to think that the enjoyment of seeing and searching can be promoted by means of coercion and a sense of duty.

Good fortune brought the friendship of Marcel Grossmann who shared his class notes, and Einstein graduated in 1900. Unable to find work as a teacher, he got a job, with the assistance of Grossmann's father, as a clerk in the Bern patent office. Apparently, the review of patent applications was not very demanding, as Einstein had the time and intellectual energy to develop several ideas that would begin the restructuring of theoretical physics. 1905 was his so-called miracle year, with publications on the theory of relativity, the famous $E = mc^2$, and the photoelectric effect, an account of the interaction between light and matter that would help launch the new science of quantum mechanics. Relativity would change the way we think about space and time; quantum mechanics would restructure the most basic understanding of matter and even the nature of cause and effect. This was not so much advancing the frontier of physics as it was starting over. It happened just a few years after Albert Michelson, an American physicist and head of the department at the University of Chicago, addressed his colleges

"... it seems probable that most of the grand underlying principles [of physics] have been firmly established and that further advances are to be sought chiefly in rigorous application of these principles ..."

But the unraveling of the grand underlying principles had begun in 1895, just 1 year after Mickelson's comments, with the discovery of something so unexpected and incomprehensible that it had to be called x-rays.

The challenge of the electromagnetic theory of light, made succinct in the equations of James Maxwell, was that the speed with which the light moves is in the law itself. That is, when you write down the equations of propagation for an electromagnetic wave, and this includes a variety of phenomena such as radio waves, x-rays, and visible light, the numerical value for the speed is there on the page. It comes right out of the properties of electricity and magnetism. This sounds pretty good, both because it eliminates the considerable trouble of having to measure the speed, and because it is always rewarding to a theorist to be able to derive a value rather than discover it or, as they say, put it in by hand. But if the numerical value of the speed is in the law, it seems impossible for the law to be invariant. Turn on a flashlight and the light moves away from you at the prescribed speed. Someone moving toward you, in a reference frame moving in the opposite direction of the beam, would measure a faster speed of the light. Speeds add or subtract from one reference frame to another, so the different frame would record a different speed of light and consequently a different law. This would be a violation of the principle of relativity.

Einstein confronted this paradox with a direct and simple solution. The speed of light is invariant, the same value in every reference frame. No matter how you are moving with respect to the source of light, or with respect to any other reference frame, you will always measure the light, any light, going from one point to another at the same speed, 3×10^8 m/s. This solves the problem—the laws of electromagnetic radiation are now invariant—but it creates other difficulties. Details of physics that had been settled would have to be reworked. For example, on the understanding that light travels as a wave, the speed of the wave would be expected to be its speed through the medium, the theorized luminiferous ether. But then the measured speed would depend on the measurer's motion relative to the medium. Einstein's solution was to simply eliminate the medium, the ether. Light, shorthand for all forms of electromagnetic radiation, is fundamentally different from all other waves. It requires no supporting medium, and, unlike any other moving thing, its speed is invariant.

This is the first of the two theories of relativity, the so-called special theory of relativity. It's not special in a good way; it's special in that it is specialized and limited, the opposite of general. This theory is limited by applying only to inertial reference frames, systems that are moving at a constant speed and in a straight line. It's fundamentally about invariants, asserting that the speed of light is absolute, the same in every reference frame, and that the laws of physics are absolute, the same in every *inertial* reference frame. It doesn't deal with the confounding effects of an accelerating frame, changing speed or direction. That will be the next step, but the special theory is remarkable progress. The blunt insistence on invariant speed of light turns out to require some fundamental changes in the Newtonian description of motion, but these can be precisely predicted and have been repeatedly tested with positive results.

The special theory of relativity preserves the anonymity of each inertial reference frame by having all the laws of physics, and hence all the experimental outcomes, the same in each frame. As Galileo pointed out, no experiment will reveal that a reference frame is moving, now carefully specified to be an inertial reference frame. This means that for descriptive purposes we can choose any reference as the one at rest. Inertial motion is purely relative because it is always possible to reference-frame-away inertial speed by choosing the reference frame in which the speed is zero. But it is not always possible to choose a reference that eliminates acceleration. In the special theory of relativity, acceleration, including rotation, is an absolute. The Earth rotates, and in the (noninertial) reference frame that is fixed on the Earth, the effects of that rotation are still evident.

The special theory of relativity not only accommodates the absolute rotation of the Earth, it suggests a novel way to measure it. It's a new kind of gyroscope. It's a gyroscope that uses light, from a theory based on the speed of light. The device was invented in 1914 by a Frenchman, Georges Sagnac, and it utilizes what has since been called the Sagnac effect. His intent was to measure the speed of light through the ether, expecting to show that, theory of relativity not withstanding, the medium of light must be there. He failed to find the ether, and, in fact, the design of the experiment was such that the outcome would not be sensitive to motion through an ether anyway, but he did lay the groundwork for a new way to detect the Earth's rotation.

At the heart of this new gyroscope is an optical interferometer. A beam of light is split by an angled, half-silvered mirror, reflecting one part of the beam off at a 90° angle while allowing the rest to transmit through. The two beams are then sent in opposite directions around a square with mirrors angled at the other three corners. They reunite where they were first separated and their respective travel-times for the round-trip are compared. If their trip times are the same, the light will recombine in phase. If one beam arrives slightly later than the other, they will be a little out of phase. Since the wavelength of light is extraordinarily small, the measurement of a phase shift is very sensitive. The smallest delay in one of the beams will result in a detectable phase shift. This is the virtue of an interferometer, a device used in a great variety of experiments and detectors of wave phenomena.

If the square arrangement of mirrors (and, in fact, any closed shape created by a series of reflections will do) is not rotating, the two beams of light have equal distances to go and with equal speeds, so they will arrive at exactly the same time—no phase shift. But if the apparatus is rotating on a turntable in, say, the counterclockwise direction, the beam sent counterclockwise will have a longer path to follow. This is because the mirrors at the corners are always moving away from the light. The speed of the light is unchanged, but it has to chase down each mirror, a moving goal post. The clockwise beam also targets a moving goal post, but in this case the mirrors are approaching and the distance is shortened. With different distances to travel, but at the same speed (and this is the key component from the special theory of relativity), the two beams will take different amounts of time to complete the circuit. When they meet back at the starting point they will be out of phase by an amount dependent on the rotation of the square.

The axis of rotation does not have to be within the square for the Sagnac effect to work. That means an interferometer fixed stationary on the ground will be affected by the rotation of the Earth. In 1925, Albert Mickelson, the same experimentalist who had declared the imminent end of new ideas in physics and who had, with Edward Morley, invented the interferometer, constructed a Sagnac device in a vacant field near Chicago with the goal of detecting the rotation of the Earth. He made it big, to enhance the effect. It's a sensitive machine, but the rotation of the Earth is a lazy one revolution per day. Using 12 inch (30 cm) water pipes and a robust pump to evacuate the air from the path of light, the round-trip was laid out as a rectangle 2,010 feet long and 1,113 feet wide (612 × 339 m). A much smaller rectangle was included to give part of the split beam a round-trip with almost no inscribed area and hence no delay due to the rotation. This was the fiducial beam, the control.

A beam of monochromatic light was split into two, each sent in an opposite direction around the huge rectangle. The phase shift in the recombined beam was compared to the control. The data from 269 trials showed a difference in travels times for clockwise and counterclockwise beams to produce a shift of 0.230 of an interference fringe, a good match with the predicted 0.236. Thus, in sending beams of light through an array of water pipes was there evidence of the rotation of the Earth.

Since then, equipment using the Sagnac effect to detect rotation has been refined and made much smaller. What is now called a ring laser is most commonly in the shape of a triangle or a loop of fiber-optic cable. Since the device functions as a gyroscope, it can detect any change in orientation in space, any absolute rotation. It makes a good compass and has been a staple in navigation systems in commercial and military airplanes for decades.

This analysis of the Sagnac effect and the description of the ring laser detecting absolute rotation, rotation in space, are done in the theoretical context of the special theory of relativity with its restriction to inertial reference frames. There's still work to be done, not just to fulfill Einstein's personal ambition of incorporating Mach's principle into the description of motion and space, but a consistent theory of gravity is missing. Newton's theory of gravity, the one we learn in science class, is not compatible with the special theory of relativity. It includes a gravitational effect that happens instantaneously between objects at any distance of separation, a violation of one of the consequences of the invariant speed of light. Nothing, neither object nor causal influence, can go faster than the speed of light. The infinite speed of an instantaneous action at a distance is impossible.

Finding a new theory of gravitation led Einstein to the second installment of his theory of relativity.

Einstein's remarkable new ideas were well received by his scientific peers, and by 1908 he had an academic position at the University of Bern. As happens to important scientists, he was lured to other institutions, first to Prague for a year, then back to his alma mater in Zurich for two, and eventually to Berlin, where he became the president of the German Physical Society. He was awarded the Nobel Prize in physics in 1921, with an acknowledgement of his multiple contributions, in particular the photoelectric effect. This began several years of travelling the world, delivering lectures and spending time in the most prestigious departments of physics. In 1933, he chose to remain at Princeton University to avoid the calamity of Nazi Germany. He remained with the Institute for Advanced Studies until his death in 1955.

During his time in Berlin, Einstein turned to the task of removing the restrictions from the special theory of relativity. Incorporating Mach's principle into the theory confronted multiple challenges. We have some explaining to do. How does relative rotation explain the Earth's equatorial bulge, Coriolis effects such as the Foucault pendulum and the eastward deflection of a falling stone, and the tenacity of a gyroscope to maintain its spatial alignment? Put another way, how does the revolution of the stars relative to the Earth cause these things?

Einstein began the general theory of relativity in the same straightforward way he started the special theory. This time, he simply removed the restriction to inertial reference frames. The more general principle of relativity makes all reference frames anonymous, even rotating frames. Just as no internal experiment can indicate steady (inertial) motion of a system, no internal experiment can indicate motion of any kind, including rotation. That's because, following Mach, there is no such thing as motion, including rotation, without reference to external things, real things like stars and galaxies.

So what are we to make of the phenomena that result from centrifugal and Coriolis forces? Einstein argued that all of these effects can be mimicked by gravity, a real force caused by real objects. As a thought-experiment, consider a closed box drifting in outer space. Now have the box accelerate, steadily speed up. Anything loose in the box will be drawn to the back, opposite the direction of the acceleration. Everything will fall back at the same accelerated rate as if there is a ubiquitous force, a force like gravity. Bingo. A so-called inertial force, the result of being in a noninertial reference frame (the accelerating box) is equivalent to the real force of gravity. Experiments done inside the box cannot tell whether the box is accelerating or is in the vicinity of a massive object generating a gravitational pull. No local experiment, that is, no experiment that measures only parameters within the system itself, will reveal that a reference frame is moving, even if the motion is noninertial. This includes rotation.

This is Einstein's version of the principle of equivalence: The effects of an accelerating reference frame are equivalent to the effects of gravity. It incorporates a similar equivalence established by Galileo that heavy things and light things fall at the same rate under the influence of gravity. Einstein's principle of equivalence was the first step toward getting relativity in line with Mach, but there was more work to be done. Gravity must act differently when there is relative acceleration between two objects than when there is no acceleration. There is a centrifugal force when there is relative rotation, but not otherwise. The details of this different mechanism were left out by Mach.

Return to Aristotle for a moment, and the idea of natural motion. In modern terms, the natural motion of a free particle, any object free of external forces, is a straight line at constant speed. In an accelerating reference frame, the box accelerating in outer space, the path of a free particle is curved. A floating stone will follow a parabolic arc to the floor; again, this is just like gravity. Invoking the equivalence principle, gravity has the effect of changing the shape of natural-motion trajectory, from straight lines to curves. The natural lines through space (and time, since this is velocity that's being affected, change of position over time) are curved. Gravity curves the natural lines of space and time.

The resulting theory of gravity, the general theory of relativity, is conceptually elegant in its bending of space and time, but the mathematics of application is complicated. There are four components to consider, three spatial dimensions and one temporal, and, since the interactions are about space and time themselves, each factor influences and alters the others. With all the cross products at work you end up with ten equations to solve for each situation. Exact solutions are possible only in idealized conditions, but this is true of the easier Newtonian law of gravity as well. The first case to be solved was that of a single object alone in the universe. How does it, according to the equations of general relativity, curve the space and time? The answer was published by Karl Schwarzschild in 1916, a year after Einstein presented the theory and 3 years before the famous test of the bending of light from distant stars as it passed by the Sun. The general theory of relativity modifies the Newtonian result of an inverse-square force. The predictions of the two theories are essentially the same when the central mass is small and gravity is weak. But results vary when the masses are huge and gravitational effects are intense. The relativistic result has the dramatic consequence of the possibility of an object so dense that when all of its mass is within a certain radius (called the Schwarzschild radius) no light can escape. This is the prediction of black holes.

Deriving this so-called Schwarzschild solution requires an added assumption that infinitely far away there will be no curvature, no influence of gravity; the space-time will be flat. This appeal to what is called a boundary condition may seem like an obvious and innocuous mathematical detail, but it turns out to be the deciding factor in whether the rotation of the Earth is absolute or relative. The flat-at-infinity boundary condition indicates that the gravitating mass does not entirely determine the shape of spacetime—and consequently the shape of natural lines and the motion of free particles—it only modifies a preexisting flatness. The flat-at-infinity boundary condition gives empty space determinate properties and thereby makes absolute rotation both defined and detectable. Mach's principle seems to have been violated.

A second exact solution to the equations of general relativity is more directly relevant to the rotation of the Earth. This is called the Kerr solution, after Roy Kerr, the New Zealander who solved the problem for a rotating black hole. He did this in 1968. Again, the flat-at-infinity boundary condition is imposed, and since the object has a determinate rotation while it is alone in the universe, the rotation is absolute. But the rotating mass creates a different geometry in the surrounding space-time, different from what is produced by a nonrotating object. It produces something called frame dragging. The lines of free-fall into the rotating mass are not radial like the spokes of a bicycle wheel; they are subtly spiraled, twisted in the direction of rotation. A gyroscope positioned near the rotating mass would tip slightly in the direction of rotation, and this is seen as a Machian effect, the influence of a real mass on the orientation of a gyroscope. But, this is still just an alteration in the antecedent flatness of the universe. The rotation of the source of gravity is determined without reference to anything else.

The general theory of relativity is catnip to a mathematical physicist, an open challenge to finding solutions to the field equations in various idealized circumstances. Kurt Gödel, a friend of Einstein with a knack for making perplexing challenges to the fundamentals of mathematics and logic, produced a solution showing that general relativity is consistent with a rotating universe. This is a little misleading. It's not that there is something outside the universe as a reference for the universal rotation. More accurately, at any location within the so-called Gödel universe that is not moving with respect to the cosmic masses there will be a Coriolis effect. A pendulum or a gyroscope will drift around even on a stand that does not move, does not rotate, with respect to the stars. It would be a Foucault pendulum on a stationary Earth. There are other bizarre features, such as the possibility of time-travel, but the Gödel universe is just a mathematical curiosity. It is manifestly not the universe we inhabit. It does, however, show that the general theory of relativity does not automatically include Mach's principle, since the mathematical formalism allows the evidence of rotation even when there is no relative motion between the masses.

There is one other idealized situation, a model of a universe, that is helpful in understanding the status of rotation in general relativity and consequently important for interpreting the evidence of the rotation of the Earth. It's a rotating shell of mass and the gravitational effects it has on the interior space. Think of the shell as representing the celestial sphere, the fixed stars, and then ask if its diurnal rotation would produce the centrifugal and Coriolis effects we observe on the Earth. Solving the equations of general relativity shows that there is a difference between the gravitational effects when the shell is rotating and when it isn't. There is frame dragging by a rotating shell, an additional component of gravity caused by the motion of the source, the massive shell. Dennis Sciama, a British cosmologist and Stephen Hawking's doctoral supervisor, called this added piece "the gravomagnetic field of the rotating universe," explicitly making the analogy to the magnetic effects that show up when electric charges are in motion. Sciama concluded, "Thus, in our theory we can regard the Earth as stationary and a Foucault pendulum as pulled around by the gravomagnetic field of the rotating universe." He awarded full-on rotational relativity to Einstein's theory.

Not everyone agreed, largely because the calculated frame dragging is not nearly enough to account for the equatorial bulge or the Foucault pendulum. Again, the relativistic effect is only a modification to a preexisting flat space. Einstein seemed to change his mind, originally claiming that the general theory of relativity fulfilled the Machian requirement of having all motion, including rotation, relative to other objects, with no reference to space itself. But by 1920 he conceded not only that rotation, unlike uniform (inertial) motion, is absolute in the general theory of relativity, but perhaps even the existence of the ether. Currently, most physicists are indifferent about the relative or absolute status of rotation in the general theory of relativity. They don't care because it makes no difference in the mathematics. Shut up and calculate. Among the few who actively work on the problem, and it is still an unresolved problem, the first question is whether the Einstein's theory is, as they say, Machian.

Most conclude that it includes some but not all of Mach's principle. Others argue that it is entirely true to Mach. The question of whether the Earth really (absolutely) rotates turns on this unresolved issue.

What would it take to make rotation a fully relative property, such that there is no difference between saying the Earth rotates or the cosmos revolves around, no difference between the old world system and the new? Within the general theory of relativity, the first step has to be eliminating the flat-at-infinity boundary condition. It has to be replaced with the conditions of masses at great distance, that is, the distribution of stuff rather than the shape of space. Then, the relative motion between the Earth and that stuff will create the gravitational field, the gravomagnetic field, to produce centrifugal and Coriolis effects. This will be some robust form of frame dragging.

Recent and ongoing work to understand frame dragging has delivered encouraging results for the relativity of rotation. The strength of the dragging of course depends on the amount of mass in the surrounding shell and its distance from the Earth. It becomes a cosmological question about the distribution of matter in the universe, a topic of active research with the suggestions of dark matter and dark energy. The conditions of perfect inertial dragging would mean that the average distribution of mass and energy in the cosmos is sufficient to keep a gyroscope aligned with the stars whether the Earth rotates or the cosmos revolves around the Earth. Remarkably, this happens if the distance to the furthest masses is equal to the Schwarzschild radius of the mass within that distance. More precisely, it's the distance to the mass and energy close enough to the Earth that light and their gravitational influence have had enough time, traveling at the speed of light, to get here in the 14 billion years since the big bang, the so-called causal distance. Do the math and this turns out to be true of the real universe only if there is a lot more mass than is accounted for in the visible objects like stars and galaxies. Allow for the possibility of dark matter and dark energy, and it turns out that perfect inertial dragging would occur if 73.7% of what makes up the universe is dark energy. This is a calculation from the principles of general relativity. Independent evidence using techniques of astrophysics estimate 73% dark energy in the universe. This seems more than a coincidence. The authors of the theoretical work,

Simon Braeck, Oyvind Gron, and Ivar Farup conclude, "Hence, the condition for perfect inertial dragging is fulfilled in our universe."

If these results are true, and they are new enough that the peer review is just beginning, then the old and new world systems are one in the same, differing only by the choice of reference frame, a choice that depends only on convenience. The evidence that proves the diurnal rotation of the Earth just as properly proves the diurnal revolution of the cosmos. Again in the words of the authors,

> All of the centrifugal and Coriolis effects observed in this reference frame can be explained as a gravitational effect of the rotating cosmic mass due to perfect inertial dragging.

It's not clear whether this is a vindication of Mach's principle. The relativity of rotation is not a matter of principle, since it does not follow automatically from the structure of the theory of relativity or the fundamental nature of space and time. It depends on the composition of the universe. It could have been otherwise.

The evidence most commonly cited for the rotation of the Earth is in the centrifugal and Coriolis effects. These are what drive the Foucault pendulum, the equatorial bulge, and the precession of a gyroscope. Effects are evidence of their cause only in the light of a dynamic theory that links the two. Focus on just one of these effects, knowing that the analysis is the same for them all. A gyroscope maintains a steady alignment with the average mass distribution of the universe; the Earth does not. This is a matter of kinematics, the basic description of what is observed, the positions and orientations of things and how they change in time. Why does the gyroscope do this? Why does the gyroscope appear to move when measured in the reference frame of the Earth, and move in exactly the same way as the fixed stars? The answer will be a matter of dynamics.

There are two alternative answers to the dynamical question. One gives the Earth absolute rotation, and thereby makes Galileo right and the Inquisition wrong. The other gives the Earth only relative rotation, and makes the disagreement go away.

On the first explanation, the alignment of the gyroscope remains fixed in space while it is carried around on the disorienting surface of the rotating Earth. The average distribution of cosmic matter, the pattern of fixed stars, is also at rest in space. This is what we mean by the *average* distribution. There is no interaction, no force between the gyroscope and the cosmic mass, but they are both in some way connected to space itself. The "in some way" has never been fully explained, but it has something to do with natural motion and inertia. The key is that some properties of space itself keep the gyroscope aligned and the stars fixed.

The second explanation is that the alignment of the gyroscope is held in place by the cosmic matter, by the stars and galaxies. There is an interaction, a causal connection between the gyroscope and the massive shell that is the rest of the universe. This is the gravomagnetic field, sometimes called dynamic gravity to distinguish it from the gravitational force that we know mutually attracts all masses in a fairly straightforward way. Here again, the details have not been fully worked out, but the conclusion is that the orientation of the gyroscope, and its apparent diurnal precession when observed in the Earth's reference frame, only reveal an orientation with respect to other objects. It's evidence of relative diurnal rotation, and that can just as accurately be described as diurnal orbit of the cosmos around a stationary Earth.

Newton advocated the first option, relying on absolute space as the reference for absolute rotation and the explanation for equatorial bulge and the eastward drift of a falling stone. Mach insisted on the second explanation. The general theory of relativity is surprisingly indifferent. The theory itself doesn't require either the absolute rotation of the first explanation or the relative rotation of the second. It's in the application of the theory that one or the other determination of rotation is made. The usual application is to start with an unaffected, flat space-time, the flat-at-infinity boundary condition, and use the theory to find out how matter alters the shape. This makes rotation absolute, with the result that the Foucault pendulum and equatorial bulge are compelling evidence that the Earth rotates. Galileo was right. But with the possibility of perfect inertial dragging, the cosmic matter provides the boundary condition and the reference for rotation. This makes rotation relative, with the result that the Foucault pendulum and equatorial bulge only reaffirm what we can already see in the sky, that the Earth and stars are in relative motion.

Remarkably, this question is still unresolved.

The Extraterrestrial Perspective

CHAPTER 14
The Final Frontier

Don't think; but look!
—**Ludwig Wittgenstein**

You can observe a lot just by watching.
—**Yogi Berra**

By the mid-twentieth century there was still no eyewitness to the rotation of the Earth. A wealth of good evidence indicated the Earth spins, but all of it was circumstantial in the sense that it was information linked to the phenomenon through some theoretical inference. This is not an unusual or unscientific situation; just the opposite, this is normal science. Even without witnesses to the interior of the Earth there is credible scientific evidence that it is solid and composed mostly of iron. The long process of evolution that produced the diversity of life on the planet is without a living witness, but again, the evidence is persuasive. Theory underwrites the interpretation of evidence, and the interpretive theories have been tested and peer-reviewed. It's a bootstrapping process, constrained by real data and honest logic.

That said, a straightforward observation, or at least a video recording, would make the case for rotation without question. Even the best theoretical support is vulnerable to reinterpretation. The history of science is a tale of changes, some small, some revolutionary, and it would be unseemly arrogance to think that we are at the end of the process, that now we've got it all right. The evidence for rotation is good, given what we know about how things move and what causes them to move, but scientists generally think that about their evidence and their foundational theories. Any lingering uncertainty about rotation, small as it might be, would be gone if we could just see the phenomenon outright.

The principle of relativity compounds the uncertainty in evidence of motion. If motion is only relative, then no local experiment can detect whether the system is moving or not. No observations within the reference frame are evidence that it is, or isn't, moving. This is why Galileo claimed that the terrestrial perspective, observations of what happens on the Earth, would never reveal the rotation of the Earth. He then broke his own rule by citing the tides as evidence of the combined action of rotation and revolution around the Sun, but the principle stands. Nonlocal observations, the celestial perspective of looking outside the system at things not on the Earth, do show that the Earth is rotating, but only relative to other things in the universe. Many who were convinced that the Earth rotates completely dismissed the evidential value of the celestial data; no doubt it showed relative motion, but just as certainly it could not show that the motion was the Earth rotating and not the celestial objects in diurnal orbit.

Now there is a new form of nonlocal evidence, measurements of the motion of objects in one reference frame from the perspective of another, the view of the Earth from outer space. It's like getting off the ship and seeing for yourself as it sails by. This is the extraterrestrial perspective. Four centuries of struggle and uncertainty since the Copernican proposal of a moving Earth have been exacerbated by the fact that we are stuck on the ship, trying to figure out if it moves by finessing the local evidence under the influence of theory. But now it's possible to just go up into space and look back, perhaps avoiding the interpretive and vulnerable theories.

The external, extraterrestrial perspective has to be at sufficient distance and for enough time to make the relatively slow rotation of the Earth detectable. It's not just a quick jump off the ground that makes you extraterrestrial. As Galileo made clear, the inertia of horizontal motion keeps a jumper in the terrestrial reference frame. The same is true for the view from a hot air balloon, since it is carried along by the atmosphere that is itself following the moving surface of the Earth. Notice that already some theoretical interpretation is creeping in as a ruling is made on what qualifies as an appropriate perspective for viewing the Earth's rotation. This is the indication that even the clear view from space might not be as straightforwardly direct observation as we might hope.

From a commercial airplane there is a clear view of the Earth, and the ground is visibly moving. But that's not the evidence of rotation we are looking for; it's evidence of the plane moving. Higher still, into outer space, the shape of the Earth and possibly its rotation become more apparent. The boundary of "outer space" is only vaguely defined, but the first photographs from space are usually credited to a 1946 project that strapped a camera to a V2 rocket liberated from Germany after World War II. The rocket flew to 105 km above the Earth and dropped the exposed film to the ground in a steel can. This was a decade before the first man-made satellite was put into orbit. Sputnik 1 had no cameras to take pictures of the Earth. Sputnik 3, however, carried a dog into orbit. There is no telling what Laika saw regarding the rotation of the Earth, and not just because she was a dog. No provision was made for safe landing on the ground. The satellite and passenger burned and disintegrated when they reentered the atmosphere.

The first man is space did safely return to the surface of the Earth. In 1961, Yuri Gagarin was the first human to orbit the planet. Presumably, at an altitude of more than 300 km, he would have seen the curved surface of the planet and watched the ground moving beneath his Vostok 1 rocket. He went once around the Earth in a little under 2 hours, making his orbit much quicker than the diurnal rotation. Rules of the international governing organization of space travel required that an official trip to space must end with the astronaut landing with the spaceship, but the Russian ship had no braking mechanism. Gagarin secretly ejected at an altitude of 7 km while his vehicle plummeted to the ground. The deception was revealed only in 1971.

As space travel became more common, photographs and videos of our own planet were shared with the public. Apollo 17 astronauts took the famous picture of the Earth from the surface of the Moon, the so-called Blue Marble image. That was in 1972. But what we need is a moving picture of the Earth rotating. The most famous and most easily accessible such video is from the spacecraft Galileo. You can find it online or watch the movie version in *An Inconvenient Truth* and see for yourself the image of the rotating Earth. Galileo was launched in 1989 on a trip to Jupiter and looked back on the way for this view of home. It's 25 hours of the rotating Earth compressed into just a few seconds of video.

Before we celebrate, the details of the Galileo footage are worth reviewing. The moving picture is in fact a composite of still photographs. NASA refers to it as an animation. This isn't really important; movies are made as a series of still pictures running in sequence fast enough to smooth the action. It's nonetheless a faithful image of the change, the motion. More important is to note that the spacecraft itself was moving when the pictures were taken. So, just as the view from a moving airplane requires compensation for the motion of the perspective, so does the view from Galileo.

The position of the camera, and its motion, are always a necessary part of the interpretation of an image. Now you can watch the images of the Earth in real time on the live-feed from the international space station. The Earth looks round, or at least the horizon is curved in a way consistent with a spherical planet, and you can see the ground rotating beneath the station. But look closely and you'll see the ground is not rotating eastward as the diurnal motion of rotation would have it. It's moving northwest. That's because the orbit of the space station is about as fast as Yuri Gagarin's, once around in 90 minutes, and what you are seeing is like what you see from an airplane, the result of the moving camera.

The importance of perspective is clearest when dealing with the video from the Himawari 8 weather satellite. The Earth appears as a colorful disk with Australia and eastern Asia easily identifiable.

Clouds move, cyclones swirl, daylight gives way to darkness, over and over. There is no rotation of the planet at all. It's a favorite reference for drive-by doubters and skeptics of rotation. Himawari 8 is, of course, a geosynchronous satellite. It is put in orbit around the Earth with a 24-hour orbital period (in contrast to the 90 minutes of the international space station or Uri Gagarin's 108 minutes), designed to stay over one point on the ground below. The Earth rotates and the satellite orbits at exactly the same angular rate. From this perspective there would be no relative motion, no imagery of rotation below.

With the newly available extraterrestrial perspective we might have hope for avoiding the necessity of theoretical interpretation of the data, that is, avoiding the appeal to dynamics, and to get by with the purely observable kinematics. But even the kinematics, the basic description of motion, relies on knowing the details of the observing conditions. The evidence is not simply a matter of pointing and saying: See, it rotates. Science is never simply a matter of pointing and saying, see. It is always informed observation, looking and thinking. Wittgenstein's "Don't think; but look!" is a false dichotomy.

The required information for the extraterrestrial perspective is the physics of how spacecraft move—rocket science. All of the images from space are from vehicles that are drifting without propulsion. Unlike an airplane that has the engines running at all times during flight, satellites, spacecraft, and the international space station are, after the initial launch and steering, without power. They are, in this sense, in a condition of natural motion. The space station floats in near-circular orbit around the Earth at a modest altitude of about 400 km. The orbit decays gradually and the station drops closer to the ground at a rate of 90 m each day, 24 km a year. The descent would continue, and even accelerate as the object entered thicker atmosphere, but for the occasional boost from visiting spacecraft. When the station is resupplied with fresh groceries and personnel, the transporting vehicle is used to nudge the main vessel back up to its proper orbit. The Galileo spacecraft experienced a sustained period of thrust in the beginning of its trip to Jupiter, this to escape the gravitational pull of the Earth, but then it was a quiet, engines-off drift with the occasional burst of small directional jets. Galileo traveled almost four billion kilometers in its 4-year adventure, using just 254 liters of fuel. That's 15 million kilometers per liter—not much use of the engines.

The videos from these spacecraft are from a moving, but not driven, perspective. They're moving fast, and to know what we're looking at and what it reveals about the Earth's rotation, we have to know how fast they're moving and in what direction. The dominating factor in these cases, as in most aspects of rocket science, is gravity.

The force of gravity between two objects depends only on their masses and the distance between them. The attractive force of gravity is what holds planets and satellites in orbit, the tether that keeps them from flying off. Any force pulling in toward the center of a curved trajectory is acting as a centripetal force, and the magnitude depends on the mass of the orbiter. Since the gravitational force also depends on the mass of the orbiter, this quantity is on both sides of the equation and it cancels. That is, gravity determines the details of orbital trajectory independent of the mass of the thing in orbit. The separation between two objects is still a factor; this is the radius of the orbit, the distance between the Sun and the Earth, for example, or between the Earth and the orbiting international space station. In figuring the orbit around a particular body with a particular mass such as the Earth, only two variables remain, the radius of the orbit and the speed of the orbiter. This is one bit of math, one equation, worth looking at because it is so simple and so informative. The orbit of anything around a central body with mass M must abide by this relation.

$$GM/r = v^2.$$

G is a universal constant, the so-called gravitational constant. For orbiting the Earth, like Yuri Gagarin, a satellite or the international space station, M is the mass of the Earth. The equation shows that at any particular orbital radius r, and consequently a particular altitude above the ground, there is only one possible orbital speed v. Everything at that altitude must have the same speed. The Moon orbits the Earth at a distance of 300,000 km. Do the math and the Moon must be moving at a speed of 1,000 m/s. If there was pea or a pebble in Earth orbit the same distance away as the Moon, it would have the same speed. Distance determines orbital speed. This is important to understanding how a camera is moving when it records video of the Earth's rotation. The magnitude of the Earth's gravitational force on the Moon is very different than the Earth's force on the pea, but the resulting acceleration is identical. It's the same principle that describes the fact that all objects fall to the ground at the same rate, regardless of mass or composition. It's the principle of equivalence.

The international space station orbits 400 km above the ground. That means the radius of the orbit is 400 km plus the radius of the Earth itself, since the center of the Earth is the center of the orbit. This fixes the speed of the station to be roughly 27,000 km/hr, much faster than the ground moves in diurnal rotation. That's why any video from this orbital altitude will display the speed of the orbiter but not the rotation of the Earth. That's what we see from the international space station.

It's worth pointing out that objects headed for Earth orbit are usually launched in an easterly direction. The United States does it from a location in Florida where the first few minutes of flight will be over the ocean such that, in the event of an early flight malfunction, the debris will fall harmlessly into the water. Cape Canaveral is at latitude 28.5°, where the surface speed of the Earth's rotation is 1,471 km/hr. Launching in the same direction as the rotation, that is, launching east, gives the spacecraft some free speed. It's not much, only about 5% of what is needed for the space station to achieve its orbit, but it's some evidence of the rotation of the Earth. The trajectory into orbit can't be due east from Florida, since the orbital radius must point straight down to the center of the Earth, the center of the gravitational force. Objects must orbit on a great circle around the center. There is only one latitude that does this, only one due-east trajectory that accommodates orbit, and that is the over the equator.

A geosynchronous satellite like the Himawari 8 has to be positioned over the equator. Again, these details are important in understanding what we are looking at when we access the video and make claims about the rotation of the Earth. A geosynchronous satellite, and there are many being used for communications, television, and monitoring the weather, must orbit the Earth in 24 hours. This prescribes a relation between its speed and the distance it travels, the circumference (and hence the radius) of its orbit. There is only one orbital distance at which this prescription is fulfilled. Every geosynchronous satellite must be 42,000 km from the center of the Earth, that is, 35,700 km from the surface, the ground. That's a very high orbit, approximately six times the radius of the Earth itself. And, given the restriction imposed by the one equation we are working with, the speed is fixed at 3 km/s, 10,800 km/hr. The perspective of the Himawari 8 is seemingly suspended high above the equator at a point over the Pacific ocean. It doesn't seem to move, since it's synchronized with that point on the Earth just below it, but it is in fact moving quite briskly. It has to be moving, and at just that speed, in order to be in orbit. Since it burns no fuel and has no propulsion, the only way it can stay up is by continuing to orbit. With this in mind, its view of a stationary Earth with no apparent rotational motion is in fact good evidence that the Earth is rotating.

The Galileo spacecraft is a little trickier, since it was not in orbit around the Earth when it took the pictures that make up the animation of a rotating Earth. Galileo was on its way to Jupiter. It's not a straight trip to cover the 600 million kilometers between the Earth's solar orbit and that of Jupiter, and, of course, it's not a matter of orbiting the Earth. The spacecraft needs to end up orbiting the Sun at the same distance as Jupiter, and it needs to achieve the corresponding speed for that solar-orbital radius. There are multiple steps involved. First Galileo had to escape the gravitational bond of the Earth. Then it used the Earth's orbital speed around the Sun, flew by Venus to use that planet's orbital speed, came back by the Earth, twice, to each time take advantage of orbital speed, and then finally on to Jupiter.

Galileo was deployed from the space shuttle Atlantis. Once separated from the shuttle, attached rockets sent the planetary probe on a course for Venus. At this point, it was going 30 km/s with respect to the Sun, the same as the Earth's orbital speed around the Sun. As it dropped toward Venus, it's kinetic energy increased, just as the speed of a falling stone increases. It would fly by Venus and return to the same Earthly distance from the Sun with the same speed, plus the little extra it picked up while close to Venus and tagging along with its motion around the Sun. Galileo was aimed to return to the Earth, now with more speed than when it departed, for a close encounter, a boost to its speed, and a turn in the direction of the Earth's orbit. It did this twice. It was on the first fly by that the video of the rotating Earth was made. The spacecraft came within 960 km of the Earth on December 8, 1990, a maneuver that added about 5 km/s to its speed. The pictures were taken over a 25-hour period December 10 to 11 when it was going 37 km/s and already 2.1 million km away from the Earth. The view is looking back on the Earth, with almost no tangential component to the motion of the camera. So, unlike the view from the international space station, there is little compensation necessary; the moving ground is in fact the rotation of the Earth.

More recent video of the rotating Earth comes from the Deep Space Climate Observatory. This is also a composite of still photographs, taken at 2-hour intervals and running for a full year. You can watch the Earth rotate, weather patterns come and go, and even the shadow of a solar eclipse that passed over our planet in 2015. The image is breathtaking, but its value as evidence that the Earth rotates is meaningless without the contextual information on the position and movement of the observatory itself. To use NASA's term, it's "parked" at a point in space between the Earth and Sun, where it stays in place without either orbiting the Earth nor the need for propulsion. It hovers at what's called a Lagrange point, named after the eighteenth century French mathematician Joseph Louis Lagrange. His name is on the Eiffel tower.

For any two massive objects like the Earth and Sun, there are five points in space at which the composite gravitational force will hold a third smaller object at rest. These are the Lagrange points. The first three were discovered by a Russian, Leonhard Euler; his name is not on the Eiffel tower. The remaining two, which are admittedly more complicated and less intuitive, were demonstrated by Lagrange in 1772. He was working on what he called "the three-body problem," a solution to the gravitational equations when there are three objects in play, not just one thing in orbit around another. Remarkably, with just three bodies or more, no exact solution is available. The dynamics of three bodies requires a series of approximations to focus in on the precise motion of the players. It was while working on this problem that Lagrange found the fourth and fifth points that bear his name.

Three of the Lagrange points are on the line that runs through the two massive bodies. These are the ones Euler pointed out. In the system of the Sun and the Earth, one of the points, L_1, is between the two. The Deep Space Climate Observatory is at L_1. This is not the point where the two gravitational forces balance, equal in strength and opposite in direction. It's the point at which the gravitational force of the Earth diminishes but doesn't cancel the force from the Sun, such that the inward pull from the Sun is exactly the centripetal force required to hold a object in solar orbit with the same orbital period as the Earth, 1 year. The Sun by itself will accommodate a 1-year orbit only at the orbital radius of the Earth. But weakening the centripetal force by counteracting with the outward attraction of the Earth, and the radius for a 1-year orbit is less. That's L_1. It's roughly 1.6 million kilometers from the Earth, in line with the Sun. So, the images from the observatory are from a vantage a little closer than from the Galileo spacecraft. It's a perspective that is neither orbiting the Earth—it's orbiting the Sun—nor moving away from the Earth. From this perspective, the rotation of the Earth is clear.

L_1 gets a lot of visitors from man-made spacecraft; it is a handy place to park. L_2 is used to position observatories, as well. L_2 is opposite the Earth from the Sun, the place where the combined centripetal forces of the Sun and Earth produce a solar orbit with a period of 1 year. The other Lagrange points are more deserted. L_3 is on the far side of the Sun, the same distance out as the Earth and orbiting at the same rate. It's where a twin Earth would be, and, of course, there has been speculation that in fact there is a hidden planet. L_4 and L_5 are not in line with the Earth and Sun. Each is at a corner of an equilateral triangle, opposite the side of the triangle that is the line between the Sun and the Earth. At this point, the composite of the two gravitational forces of the Sun and the Earth points to the center of mass between the two. L_4 and L_5 orbit this point, as do the other three Lagrange points.

The Deep Space Climate Observatory is parked at L_1, but that doesn't mean it's stationary; it's orbiting the Sun. The physics of gravity indicates that it is not orbiting the Earth, and this makes it a useful perspective from which to observe the rotation of the Earth.

Knowing the context and the relevant science gives meaning and credibility to the evidence.

All of these analyses of orbital mechanics have been done using the Newtonian representation of gravity and Newtonian laws of motion. They are also done with no regard for the rotation of the source of gravity. That is, in figuring out the force of gravity, the required centripetal force to hold something in orbit, and the resulting mathematical relation between the radius and speed of an orbit, there was no concern about the rotation of the central massive object. The Sun rotates, but we didn't factor that in when calculating solar orbits. We didn't need to know whether the Earth rotates or not to pin down the details of satellite orbits. This means that the interpretation of the extraterrestrial perspective on the Earth's rotation is independent of the hypothesis that the Earth rotates. That's good science.

But, the interpretation is not independent of the fundamental understanding of gravity. Change the gravitational theory to general relativity, and frame dragging shows up. In this theoretical context, orbital dynamics are not uninfluenced by the rotation of the central object. Add Mach's principle and the possibility of perfect inertial dragging, and every image from any vantage can be described as an Earth that does not rotate. The Sun and L_1 are being dragged around the Earth by the dynamic gravitational force of the orbiting cosmic masses. On this interpretation, it's not that the Earth does not rotate; it's that the situation can be described either way with equal accuracy. All motion is relative.

While we're up in space we should take advantage of recent satellite measurements of the Earth's magnetic field. This will be an extraterrestrial perspective on a subterranean phenomenon that is related to rotation. William Gilbert speculated that magnetism was the driving force of the Earth's rotation. Kepler extended the magnetic influence and used it, or something like it, to couple the rotation of the Sun to the orbits of the planets.

A lot has been learned about the Earth's magnetic field since the seventeenth century. It's not just the orientation and patterns of the field lines, but some surprising, and almost alarming, trends have been discovered. For example, the strength of the field is decreasing, roughly 10% since careful measurements in 1832. Paleomagnetic data indicate that the polarity of the planet has flipped, north to south, on several occasions. The timing is erratic, from as little as 10,000 years between flips to as long as 25 million years. The process of reversal takes about 5,000 years. There is some speculation that the current decreasing strength could be prelude to a polarity reversal. The ability to monitor magnetism from space has facilitated a more precise focus on the fine-structure of the Earth's field. The European Space Agency launched its Swarm mission in 2013, three identical satellites orbiting in formation. NASA followed in 2015 with the Magnetic Multiscale mission, this one with four identical satellites in a tetrahedral pattern. The multiple measuring devices provide a three-dimensional image of the magnetic field.

Even with the sophisticated data-gathering and enhanced imagery, there remains some uncertainty on the details of the causal connection between the Earth's rotation and the magnetic field. The fundamentals of magnetism are much clearer than they were in the seventeenth century, but the interior workings of the Earth are still somewhat mysterious. Moving electric charges induce a magnetic field. This is the basis of the analogy alluded to by Mach, that an additional interactive force arises when there is relative motion of the source—it works with an electric charge, so it must work with a gravitational charge, a mass, as well. Rotation is a kind of movement, so a rotating electric charge would result in a magnetic field. Gilbert was on to something, but he got the causal relation backward. It's not that magnetism causes the Earth to spin; rather, the rotation of the Earth is the cause of it's magnetism. More accurately, the rotation is part of the cause, one of several contributing factors, each necessary but none individually sufficient, to sustain the Earth's magnetic field. It's a complicated interaction, but even though rotation is not the only contributing factor, if it's *necessary* for there to be a magnetic field, that's good news for using magnetism as evidence for rotation. If there can be no magnetism without rotation, then any measurement of a magnetic field will be a sure indication that the Earth rotates. Mach's principle, however, will require us to confront the question of whether this is rotation in the absolute sense or merely relative to other things in the universe.

The current model of the Earth's magnetic field indicates that it is not a permanent magnet, that is, not a condition that was formed in the past and simply remains. It's not like the equatorial bulge, formed when the Earth was more malleable and now frozen in a shape that would persist even if the Earth was not rotating. The magnetic field has been part of the Earth for at least three billion years, but the high temperatures of the interior would not sustain magnetism for more than 15,000 years. So, there must be an ongoing sustenance for the field.

The basic dynamo theory of the Earth's magnetic field was proposed in 1919; the details are still being worked out. The term comes by analogy to an electric generator, a dynamo that converts kinetic energy into electricity. A dynamo involves magnets, motion, and electric charges. Following this model, three conditions are prerequisite for a planetary magnetic field: a fluid with unbound electric charges that can conduct an electric current, movement of the fluid, and a preexisting magnetic field through which the fluid moves. The outer core of the Earth is liquid iron, a suitable conductor of electricity. Heat from deep within the planet causes convection in the liquid, and a Coriolis force, the result of rotation, organizes the convective currents into helical columns. A subtle but persistent magnetic field from the Sun provides the background, the seed field. The electric charges in the molten iron move through the preexisting magnetic field and convert the kinetic energy of motion into an enhanced strength of magnetism. Without the movement the field would quickly decay and disappear.

The mathematical representation of this process is outrageously complicated. It involves the interaction of electrodynamics, thermodynamics, fluid dynamics, and gravity. The mathematical modeling requires ten interdependent equations to be solved simultaneously. You may recall the challenge and frustration of solving just two simultaneous equations, and that was algebra; these are partial differential equations. Solutions to this set of equations require knowing the boundary conditions, numerical details on things like temperatures and pressures in the interior of the Earth, viscosity of the materials, depths of the interfaces, and so on. These are not well-known, and all of this adds up to uncertainty.

Computer simulations directed by the math have produced credible results. Not only do they show the generation and persistence of the magnetic field, they are consistent with the possibility of reversing polarity. The inner core of the Earth is solid. It's hot and mostly iron, but high pressure prevents the iron from liquefying. The solid center resists changes such as the flipping of magnetic poles, and without that resistance the reversals would be more frequent and more regular. The Sun's magnetic field reverses polarity regularly every 11 years. The Sun has no solid core. The computer modeling of the dynamo of magnetism also predicts that the inner core of the Earth rotates a little faster than the rest of the planet. This so-called super-rotation is minimal, no more than half a degree per year, but there is independent evidence that it happens, evidence using the analysis of seismic waves. There are a lot of moving parts in the generation of the Earth's magnetic field, but they seem to be fitting together in the most recent scientific description. One of those pieces is the whole Earth itself, rotating.

Again, the key is that rotation is a necessary condition of the magnetic field. This seems corroborated by what goes on in other planets. Venus has an iron core but no measurable magnetic field. It's rotation is almost nonexistent, once around in 243 days. Little rotation is correlated with little magnetism. This is consistent with rotation being necessary. The weak link in the inference is the limited knowledge of the consistency of the planet's core; it could be solid iron. If so, that would eliminate another well-known necessary condition, the fluid conductor. Jupiter provides the opposite kind of evidence. It has a very brisk rotation, taking only 10 hours for the huge planet to spin once around. Jupiter has a strong magnetic field. The composition of Jupiter is mostly hydrogen, not normally a conductor of electricity. But at the center of the planet the hydrogen is in an unusual metallic state. It's too hot to be a liquid, but under too much pressure to be a gas. The result is a disassociation of nuclei and electrons, a churning vat of loose protons and electrons that flow like a liquid and conduct electricity like a metal. The conditions are perfect for the dynamo generation of a planetary magnetic field.

There is an abundance of good evidence for rotation to be found in the fact that the Earth sustains a magnetic field. The logic is good: If there is a magnetic field, there must be rotation. And the supporting interpretive theories are well established, the electrodynamics, thermodynamics, and so on. But again, Mach. All of the rotational contribution to the magnetic field is in the influence of Coriolis forces. If there is perfect inertial dragging by the distribution of cosmic mass, then this is not absolute rotation. It can be described, quoting Sciama again, as "the gravomagnetic field of the rotating universe." The dynamo depends on the relative motion between Earth and cosmic mass, making the magnetic field good evidence of relative rotation. But we never needed evidence for that, since it is plainly visible in the kinematics of celestial motion.

There is a spectrum of indirectness in the evidence for rotation as gathered from space. We see for ourselves the video—the animation—as relayed from spacecraft. That's pretty direct, almost an eye witness. At the opposite end of the spectrum is the claim that a magnetic field is evidence of rotation. That invokes a lot of interpretive theory and parameters, some of it poorly understood. But in both cases, and all cases in between, some amount of interpretation, reliance on background knowledge, is required. That's how science works. It doesn't rest on naïve experience; observations and evidence are endorsed and understood in the context of the stable network of scientific beliefs about nature. Whether it's complicated theories about magnetism or the more basic appraisal of the context and conditions of observation, no image speaks for itself. Recall Ptolemy's warning against "making the mistake of judging on the basis of [one's] own experience instead of taking into account the peculiar nature of the universe."

Yogi Berra may have been right, "you can observe a lot just by watching." But to know what it is you are observing, and to use it to expand your knowledge to what you can't observe, it will take more than just watching. There is more to consider.

CHAPTER 15
All Things Considered

The truth for which Galileo had suffered remains the truth, although it has not altogether the same meaning as for the vulgar, and its true meaning is much more subtle, more profound and more rich.
—**Henri Poincaré**

The history of evidence for some natural phenomenon is often a concurrent development of both the clarification of the idea and the empirical reasons to believe it's true. There is some give and take between what it means to say that nature has some particular quality and the evidence that shows that it does. You might not expect this sort of conversation and readjustment on an issue so straightforward as the rotation of the Earth, but it is a fair question to ask: What does it mean to say the Earth rotates? It is probably worth asking before, or at least along with, that driving question for our work: How do we know the Earth rotates? You would get an answer from your science teacher on the how-do-we-know question, but there would probably be just a quizzical look on the what-does-it-mean. What part of "the Earth rotates" don't you understand? It's the ambiguity between absolute and relative rotation that needs to be clarified, an ambiguity that facilitates equivocation. We need to figure out what we're looking for, and what it looks like, in order to fairly decide if we've found it.

In the beginning, that is, in the ancient Greek discussions of cosmology and the status of the Earth, the Earth moving or not was a simple concept. It went without saying that the motion in question was absolute and that there was a determinate sense to saying simply that the Earth rotates or it doesn't. This was before any explicit doubts or defense of the existence of a universal container in which the stars and planets and the Earth are situated. Each individual constituent of the universe is at a particular place in the container, and some of them move around. Some scientists argued that there was no way to know whether something was moving in this absolute sense, and anything that can be observed only reveals movement relative to other things. The inability may have been specific to observations of celestial phenomena or terrestrial. But this just meant that the reality was hidden, not that absolute motion was unreal. There is a simple fact of the matter whether the Earth rotates or not, even if there is no available evidence and no simple way to prove it.

Skip ahead to the present; the understanding of what it means to say the Earth rotates is unchanged since antiquity. At least in the classroom and in the public understanding of our own planet there is an implicit assumption that it's absolute rotation we're talking about. We automatically interpret Galileo's muttering, and yet it moves, to mean moves through space, sharing with the ancient Greeks a presumption of a stable cosmic stage on which events unfold and in which things are positioned and moved. It goes without saying; it even goes without thinking about. And yet, when the details unfold there is often a clarification of the difference between a solar day and a sidereal day. That is, the rate of rotation of the Earth is acknowledged to be relative to some other celestial object of reference. The Earth rotates more slowly with respect to the Sun than the distribution of stars. This is a first step to relativity. If the rate of rotation depends on the reference, it must be possible to choose a reference that results in zero rotation. Without explicitly explaining whether it is relative or absolute rotation that we are talking about, the modern conversation harbors an embedded ambiguity.

Between the ancient accounts and the present, and in the more technical analyses today, there is better clarity on the meaning of rotation. Newton was explicit. Absolute rotation is real and demonstrable, and the Earth exhibits the symptoms. The spinning-bucket experiment shows not only that absolute rotation is real but that it can be detected. Absolute space, though entirely unobservable itself, must exist, as it provides the only explanation of the behavior of the water in the bucket. And the Earth is spinning in space, as seen in the evidence such as the equatorial bulge.

The postmodern deconstruction of the Newtonian demonstration of absolute rotation, Mach's analysis of the spinning bucket, forced a confrontation between the intuitive idea of motion in space and the perplexing idea of there being no enduring backdrop of cosmic events, only the things themselves, actors on no stage at all. Dealing with the distinction between relative and absolute motion is generally unseen except by a small group of physicists and philosophers concerned with the most fundamental nature of space and time and motion. It's hidden and dismissed as of no practical consequence. Day-to-day physics and textbook physics don't depend on the difference, but clarity on the claim that the Earth rotates does. What, exactly, does it mean to say that the Earth rotates?

It helps to point out that rotation is what can be called a two-place property. Strictly speaking, a description of something rotating is incomplete without the reference (perhaps implicit) for the motion. It's rotating with respect to something else. We deal with two-place properties all the time. The property of being taller is a good example. The proposition "A is taller" is incomplete, obviously so. It needs the reference before it is meaningful and before it can be either true or false. "A is taller than B." Now this makes sense, and it automatically allows for multiple descriptions of the one situation. "A is taller than B." "B is shorter than A." They say the same thing, and they are true (or false) under exactly the same circumstances.

Apparently, "day" is another two-place concept, incomplete without the reference (almost always implicit) for the demarcation. To say "It has been one full day" is not meaningful, and it is neither true nor false, until the specification of solar day or sidereal day, or some other, less common, designation such as lunar day. There could even be an absolute-space day, one full rotation with respect to space itself. Presumably its duration would be nearly equal to the sidereal day, since the display of stars is more or less fixed in space. But it could never be directly measured, since the reference, space itself, is in no way observable. There is no practical reason for considering alternatives to the more common solar day, and, if you are an active astronomer, sidereal day, but the possibility demonstrates the implicit requirement of a reference.

We say that the Earth rotates; we also say that the Earth is round. The shape is a one-place property; it is completely described without reference to anything else. This is part of the reason there has been both clear understanding and little disagreement about the roundness of the Earth.

When no reference of rotation is given, the default is space itself. This was made explicitly by Newton, and explicitly challenged by Mach and, before him, Gottlieb Leibniz and George Berkeley. They brought to light the metaphysical implication of absolute rotation, the existence of an unobservable reference.

No one has claimed that space itself can be seen. Consequently, no one can claim that absolute rotation can be directly observed— the second piece of the two-place relation is hidden. More generally, no absolute motion is observable, not the Earth's rotation or the rotation of anything else, whether we're on the thing or off, whether the data are local or distant. No absolute motion of any kind, inertial or noninertial is observable. There can be no eyewitness to the phenomenon of motion relative to space itself. This doesn't mean there is no such thing, only that it can never be seen. It leaves open the possibility of indirect evidence.

Absolute inertial motion, straight and with uniform speed, is hidden even from indirect evidence. Nothing can detect a steady movement with respect to absolute space. This is the principle of relativity. Galileo put it on the books and used it to argue that no terrestrial data would reveal the rotation of the Earth. He was considering rotation as a natural, steady motion, and assuming that it was absolute motion at issue. It was Newton's revision of the concept of natural motion, and consequently the distinction between inertial and noninertial motion, that classified rotation as forced. This reopened the possibility of indirect evidence.

Whether or not there can be evidence of absolute noninertial motion depends on the theory of dynamics applied to the interpretation of the data. And, or course, theories are subject to challenge and change. That's science. Appealing to the existence of absolute space allowed Newton to make sense of the shape of the water in the spinning bucket. It provided an explanation. But it wasn't the only possible explanation, and this is the general plight of scientific explanation. We might argue that Newton's is the best explanation, and hence the most likely to be true. This, too, is a very common inference in the life of a scientific explanation. But what counts as *the best* explanation? What are the criteria that make one explanation better than another, and, importantly, how do those criteria indicate a greater likelihood of truth? Simplicity, often marketed as Occam's razor, is the go-to standard for a good explanation. The account of things that limits the presumptions and moving parts is the one more likely to be true. Other criteria for quality of explanation may import metaphysical ideals such as the affirmation (or denial) of crystalline spheres or natural motion. There could be methodological guidelines, for example, Mach's robust empiricism that prohibits the appeal to anything that cannot be observed, a move that allows the epistemic quality of things to dictate their metaphysical status. With all of these possibilities for the ruling on the best explanation, the appeal is to nonempirical properties—an ironic situation that did not go unnoticed by critics of Mach-style empiricism.

Much of the history of debate about the rotation of the Earth has been obscured by the failure to confront the difference between relative and absolute rotation. We can make progress toward fixing that by using the modern template of distinguishing between kinematics and dynamics, and looking more closely at those details. The study of motion is divided into two aspects, the descriptive and the explanatory. Kinematics describes the movement of objects; dynamics explains it by giving the details of the cause. Kinematics is about position and the change of position, velocity, and acceleration. All of these kinematic properties are observable, given a suitable reference frame and identifiable reference objects. It may require instrumental enhancement, using a telescope, for example, to record the movements of the moons of Jupiter, or a microscope to precisely measure the precession of a gyroscope, but this is not the sort of theoretical interpretation that shifts the information from being observation to being indirect evidence. The data of stellar aberration and parallax are kinematic. They are agreeable across theoretical commitments, discounting the pig-headed denial of refusing to even look through the telescope.

Dynamics gets to the cause of the kinematics effects. It is somewhat anachronistic to describe Aristotelian dynamics, but insofar as he cited causes of motion they were generally teleological, in terms of the goal and proper place of an object of motion, where it was headed rather than what was pushing or pulling it. This often involved a particular place or direction, that is, an aspect of space rather than the influence of another object. In this way, Aristotelian dynamics presupposed a preexisting, real, independent space. Copernicus, it's fair to say, offered no dynamical theory at all. He was a pure-hearted astronomer, interested in description with no dabbling in the mechanical forces at work. With no replacement of the Aristotelian physics it was challenging to interpret the kinematic astronomical data in a new way. Kepler offered a speculative and somewhat mystical dynamics for the cosmos, employing magnetism to make things move as observations indicated. Galileo, despite his monumental contributions to terrestrial physics, provided no new dynamical theory for astronomy. The principle of relativity and its companion concept of inertia are about the nature of motion, what happens and what it can and cannot reveal about the movement of the system, with no explanation of why things move the way they do. This is the kinematic prelude to Newton's first law of motion, the first law of dynamics. Newton was the first with a clear and explicit theory of dynamics. The second law, $F = ma$, makes the explicit and quantitative connection between motion (the acceleration a) and its cause (the net force F). This is the link between kinematics and dynamics still used for day-to-day physics.

In terms of kinematics and dynamics we can make blunt and unequivocal assessment of the two big ideas that got us started. First consider the kinematics.

The Earth rotates. In the context of kinematics, this is true, but so is the statement that the cosmos revolves around the Earth once a day. Describing the situation one way rather than the other only reflects a choice of reference frame. It's the same logical relation as saying A is taller than B, admitting the equally accurate alternative perspective by saying that B is shorter than A. There is just one relationship there, described in two ways. Similarly, there is just one universe, described by two world systems, so-called by Galileo. The heliocentric model in which the Earth moves and the geocentric model in which it doesn't are kinematically equivalent.

All motion is relative. In the context of kinematics this is true. Whether it is inertial or noninertial, in a straight line or around in a circle, only relative motion can be observed and described. Kinematics accommodates full-on relativity.

Turn on the dynamics, and the two ideas must be reevaluated.

All motion is relative. There is a fact of the matter in this; the statement is determinately true or false, but there is uncertainty which it is. Newton said decisively that it's false. The general theory of relativity, the current state of the art in space-time physics, is unclear on the issue, although it seems to be siding with Newton. Mach was as clear as Newton but with the opposite response; the statement is true. Verging on violation of his own strict empiricism, to admit no unobservable metaphysics into any scientific description of nature, he used the nonexistence of absolute space as the unquestionable foundation for the development of dynamics, even when it required new (unobservable) forces at work to explain such phenomena as centrifugal and Coriolis forces. Mach's insistence on the relativity of all motion was a matter of principle, but recently there is the possibility that it is true, but only because the universe happens to have just the right amount of mass to produce the perfect inertial dragging that provides the revolving cosmic matter sufficient dynamic influence to account for centrifugal and Coriolis forces. It comes as a surprise that such a fundamental truth about the nature of space and time, whether there is or isn't absolute space and consequently absolute rotation, is contingent on the contents and distribution of stuff.

The Earth rotates. In the context of dynamics, this is complicated by the uncertainty about the relativity of rotation. Newtonian dynamics distinguishes between the two world systems. Specifically, the geocentric model is dynamically impossible, given what we know about the forces required for orbit and the nature of gravity. The Tychonic model that has the Sun orbiting the Earth and leading the other planets around with it is not only outlandish and unnatural, it's a violation of laws of physics, that is, laws of Newtonian dynamics.

A decisive judgment on the rotation of the Earth requires adopting one dynamical theory or another to interpret the observed data. Kinematics is observable; dynamics is not. And all the kinematic evidence in the world cannot determine the truth of any particular dynamical theory. Observed effects—the way things move relative to other things—cannot unequivocally demonstrate the unobservable cause—the forces at work. Kinematics, as the philosophers would put it, underdetermines dynamics.

In science, as in life, we have to make decisions about things that we can't directly observe. That doesn't mean we're left with nothing but faith, or guessing, or dogma. There is a lot of room to operate between the extremes of certainty and ignorance, and that's where science flourishes. It is somewhat misleading to offer as a clean dichotomy the difference between direct observation and indirect evidence, the eye witness and the circumstantial. There is a difference, but it comes in degrees. Credible, admissible, observation must include an account of the proper conditions for viewing. There is that amount of interpretation in even an eyewitness account. More important, the indirectness of evidence is a matter of degree, both in the amount of interpretation involved and its quality. Evidence may require more or less theoretical background, from the fairly simple equatorial bulge indicating an active force to the almost overwhelmingly complex connection between the Coriolis effect of rotation, in a crowded party with other forces, giving the Earth its magnetic field. Both cases, indeed all cases, of interpreted evidence depend on theoretical background, but clearly some depend more than others. There is also the consideration of the status of the theories used to consider in evaluating the credibility of evidence. Well-tested interpretive theories make the inference from data to evidence more reliable than a use of speculative or outright suspect theories. And independent theories, those not beholden to the evidence in question prevent a self-serving circularity in the inter-pretive process. All of these criteria have intuitive analogies in court-room examples. Does the witness stand to profit from the conviction or acquittal of the defendant? Is the witness a reliable authority? Is the forensic technique well established and well tested? And so on. All of these concerns are raised to put circumstantial evidence in perspective and properly on the spectrum from credible to dubious.

Back to science and the rotation of the Earth. As more data are available, and theories are required to not only explain what is observed but also fit coherently with each other, there is increased good reason to believe the whole package is accurate—evidence, interpretive theories, and the larger theoretical network. But sci-ence keeps going, and things do change, sometimes abruptly and sometimes in an almost wholesale way. There is no reason to think that our place in the history of science is unique, any more than to think that our place in the physical universe is unique. We are not at the center of the world, and we are not at the end of scientific change. What cosmologists proudly call the Copernican principle, the humbling realization that we do not occupy a special place in the universe, applies to both our position in space—we are not at the center of the universe—and our place in time. Our moment is not unique. Things change, and it's good to keep that in mind when doing science or history. From some future perspective, ours may be the old world system.

We evaluate the worth of our own evidence in light of the current scientific understanding, the theories we now accept as describing nature. We should evaluate evidence used at other stages in the history of science in light of the scientific understanding at those times. Being reasonable and being scientific entail judging the credibility of evidence all things considered, that is, all information available at the time considered. In this light, ancient Greeks, with the implicit assumption of the reality of absolute motion and a dynamical understanding that did not include inertia, were quite reasonable in concluding that the available evidence, celestial, and terrestrial, indicated the Earth does not move. At the time of Copernicus and Galileo, still invoking an unacknowledged, and perhaps unaware, belief in absolute rotation, but with no clear dynamical laws, the kinematic data and conceptual criteria such as simplicity and elegance in the model gave some reason to believe the Earth does move. It was not overwhelming proof, and resistance to changing to a new world system was not unreasonable. Newton provided both the explicit assumption of absolute rotation and the dynamical context to give the terrestrial evidence the authority for scientists to reasonably declare that the Earth rotates. This is the context still at work in science classrooms and news sources like NASA and reader-friendly books and magazines. Looking deeper into the physics, general relativity, the current standard in describing gravity and cosmic phenomena, makes it complicated and unclear whether it is reasonable to believe the Earth rotates. Despite the name of the theory, there seems to be a lingering need for absolute space. General relativity, and the subtle evidence in its testing, do show the reality and effectiveness of frame dragging, and this provides the mechanism and the possibility that, with enough cosmic mass, the rotation is simply relative. The conclusion is unresolved.

Mach's principle is still an unfulfilled promise, and you might say a half-baked dynamics. Newtonian dynamics works and general relativity is well tested, and both are complete in mathematical detail. In this context, our place in the history of science, it is reasonable to say that the Earth rotates, absolutely. We know the Earth rotates, but not because we have seen it rotate. We know it because we understand the dynamics of nature, what forces are at work and what observable effects they have. We are following the advice of Ptolemy, "taking into account the peculiar nature of the universe." To say we know the Earth rotates is not to claim that we are certain about it. There is consensus now, as there was consensus in ancient and medieval times that the Earth stands still.

By following the evidence for the rotation of the Earth in its historical context, we can ask fair questions about the good reasons to accept or reject the hypothesis, all things (at the time) considered. There are two ways to put the question. When and how did it become reasonable to believe that the Earth rotates? Alternatively, when and why did it become unreasonable to maintain that the Earth stands still? In light of the lingering uncertainty about relative or absolute rotation, the second version of the question allows a more direct answer than the first. Since the discovery of dynamics, it is unreasonable to claim that the Earth stands still, at least in the absolute sense. The evidence in hand, things like the equatorial bulge, Coriolis effects, and now the view from space, justifies the claim that the Earth rotates, absolutely. Alternative (and tentative) dynamics reinterprets all that evidence as indicating that the Earth rotates but only relative to other things, the fixed stars, meaning that the Earth rotating is one of two equally accurate descriptions of the cosmological situation. One way or another, absolute or relative, the Earth rotates. Denying rotation is denying the evidence. It is unreasonable to claim that the Earth is absolutely at rest.

We know that Galileo was right; the Earth does move, one way or another, relative or absolute. More decisively, we know the Earth does not stand absolutely still. The evidence is clear.

The Catholic church has come to terms with this empirical reality. Despite scripture, official church actions and documents have reversed the condemnation of the heliocentric world system and no longer prosecute or suppress declarations that the Earth is in motion. It has been an incremental process that has never explicitly endorsed the Copernican model; it has only released it, and its advocates, from ecclesiastical censure. In 1741, 200 years after the publication of *On the Revolutions of the Heavenly Spheres*, but only 5 years after the measurements indicating that the Earth bulges around the equator, the Catholic church allowed the publication of a censored version of Galileo's *Dialogue*. The text was edited to describe the Copernican ideas as hypothetical, and it was supplemented with the full statement of Galileo's condemnation, that is, the verdict of his trial before the Inquisition, and his own declaration that he never believed the Earth moves. In 1758, a general allowance of publications of heliocentric ideas was issued, but without explicitly removing from censure the specific works of either Copernicus or the unedited version of Galileo's *Dialogue*. This restriction wasn't lifted until 1822 with a declaration by the College of Cardinals.

> The printing and publication of works treating of the motion of the earth and the stability of the sun, in accordance with the opinion of modern astronomers, is permitted at Rome.

Still, the original versions of both Galileo and Copernicus remained on the church's index of prohibited books, the *Index librorum prohibitorum*. They were finally cleared of this prohibition in 1835 when the next edition of the list was published without including either. The Catholic church finally discontinued publication of the index entirely in 1966. The collected works of Giordano Bruno were on to the end. Remarkably, Charles Darwin's name never appeared.

This progressive allowance of heliocentric publications never quite admitted the truth of the heliocentric model of the solar system; it doesn't say that the Earth moves. It allows for discussion of the idea and it removes the explicit rejection of the rotation of the Earth. There is no admission of guilt or negligence or mistreatment of the idea; the evidence accumulated and the church's consideration of the theory evolved.

The church has also reconsidered the status of at least one of the advocates of the Copernican system, Galileo. In 1979, Pope John Paul II suggested an official review of the famous trial. As with the prohibition of books, this was not to be about the scientific matter of whether or not the Earth moves, but only about the the process of condemning the man. It would be about the trial but not the verdict. A commission was formed in 1981, and after 11 years of study, Galileo was officially pardoned, or, by the word of the church, rehabilitated. This made headlines in our scientific culture, drawing attention to the amount of time it took between conviction and exoneration. No one put it quite this way, but the Earth turned fully around 131,243 times between the condemnation of Galileo on June 6, 1633, and his rehabilitation on October 31, 1992. That's the number of days in those 359 years, the number of complete rotations with respect to the Sun. It's a different number if you use the stars as the reference. And it's simply zero if you use the Earth itself.

The pope's own summary of the conclusion offered no suggestion that the Earth might be in motion. As he said, "The Bible does not concern itself with the details of the physical world." The report and rehabilitation are about the procedures of the trial and the methods of science. The verdict was set aside, that is, the scientist should no longer be condemned, but not because his claims were true but because the methods of science cannot be judged by the standards of religion. But the pope seemed to blame Galileo for the unfortunate outcome of the trial. "He rejected the suggestion made to him to present the Copernican system as a hypothesis, inasmuch as it had not been confirmed by irrefutable proof." This, John Paul pointed out, would be a violation of Galileo's own scientific method that is clear on the uncertainty of scientific results. The pope prevaricated; the transcript of the trial, as well as Galileo's published writing on the Copernican world system, included admissions of uncertainty. From Galileo's deposition

> In regard to my writing of the Dialogue already published, I did not do so because I held Copernicus's opinion to be true. Instead, deeming only to be doing a beneficial service, I explained the physical and astronomical reasons that can be advanced for one side and for the other; I tried to show that none of these, neither those in favor of this opinion or that, had the strength of a conclusive proof and that therefore to proceed with certainty one had to resort to the determination of more subtle doctrines, as one can see in many places in the Dialogue.

Even presenting the heliocentric model as a tentative hypothesis was a punishable offense. Galileo was officially told, both before the trial and in the written condemnation, regarding the idea that the Earth moves and is not at the center of the universe, that he was "not to defend it, nor even to discuss it." Giving evidence, as you would for a hypothesis, would be seen as defending the idea. It would certainly entail discussing it. It's not the hubris of certainty that got Galileo in trouble; it was defending the Copernican system as a hypothesis.

There is another meaning of "hypothesis" that was current at the time of the trial and often used in reference to cosmological models. Ptolemy in some places described his system of epicycles and deferents as a hypothesis, but he didn't mean it was unproven and tentative. He meant it was simply a mathematical device for keeping track of the planets, neither true nor false, merely useful. A hypothesis was no more than a way to save the phenomena. This was the ploy by Osiander in his uninvited and unendorsed preface to Copernicus' *On the Revolutions*, a reduced ambition, and consequently less of a threat, from the heliocentric model. Galileo did ignore this sense of hypothesis, clearly assuming that one or the other of the world systems is true, but this is not the sense of hypothesis the pope was talking about in 1992.

Robert Bellarmine, the church official who first put in writing to Galileo the admonition not to defend or teach the heliocentric world system, confided in a letter to Paolo Antonio Foscarini, a theologian with Copernican tendencies, that if there was irrefutable proof that the Earth moves, scripture would require reinterpretation.

> If there were a real proof that the Sun is in the centre of the universe, that the Earth is in the third sphere, and that the Sun does not go round the Earth but the Earth round the Sun, then we should have to proceed with great circumspection in explaining passages of Scripture, which appear to teach contrary, and we should rather have to say that we did not understand them than declare an opinion false which has been proved to be true. But I do not think there is any such proof since none has been shown to me.

Bellarmine put the burden of proof entirely on the new world system and set the threshold, a "real proof," impossibly high, or at least scientifically impossible. This is the standard that was applied during the trial, and the mismatch between certainty and scientific method that we now recognize became the church's reason to release Galileo from condemnation. Thus was Galileo rehabilitated, his legacy and spirit were allowed back in the church in 1992.

Galileo's body had been allowed back in the church much earlier. He occupies a monumental tomb in the Basilica of Santa Croce in Florence, neighbor to both Michelangelo and Machiavelli, just across from Dante's empty sarcophagus. This was the initiative of Vincenzo Viviani, a student and assistant to Galileo. He was the one who had noted the bothersome tendency of a pendulum to slowly drift around in a westward direction and put a stop to it by constraining the swing with a second string. He had noticed this Coriolis effect, later amplified in a Foucault pendulum, but missed the importance as evidence that the Earth rotates. Without the interpretive help of Newtonian dynamics, the informative signal was lost in the noise. As a biographer of Galileo, Viviani was also responsible for the apocryphal tale of his teacher atop the leaning tower of Pisa to drop two balls, one heavy and the other light, to demonstrate the equal rate of falling.

Viviani was 20 years old in 1642 when Galileo died and was unceremoniously interred in a small and undignified chamber under the bell tower of the Basilica of Santa Croce. Sixty-one years later, after a productive scientific career, the student joined his teacher in the same grave, but with provisions in his will to construct the more appropriately magnificent and public resting place for his mentor. Among his motives was the hope that reconciliation between church and the outspoken Copernican would ease the tension and allow Galileo's scientific legacy to flourish.

The tomb near the front entrance of the cathedral was finally ready in 1737 and Galileo's body—most of it—was exhumed and placed into its sepulcher. At the event of disinterment, three fingers, a tooth, and at least one vertebra were removed from the corpse and kept as relics. This was not at all an uncommon practice at the time, to appropriate pieces of dead scientists, like saints, as souvenirs and reminders of their contributions to the understanding of the physical world.

René Descartes, whose *Meditations on First Philosophy* was entered onto the *Index librorum prohibitorum* in the same year as Galileo's conviction by the Inquisition, died in 1650 and was buried in the icy ground of Stockholm where he had been employed as tutor to the young Queen Christina. His body was exhumed in 1666 to be transported back to his native France. The right index finger was detached and gifted to the French ambassador, while the remainder was boxed for shipment. It arrived without the head, apparently removed as a necessity for fitting the corpse into a small box. It's an ironic loss, since Descartes is famous (or notorious) to philosophers for proposing a fundamental distinction between a person's mind and body, between the intellectual and corporeal self. His skull was eventually recovered, or at least a skull showed up and it was officially declared to be Cartesian, but it was not reunited with the rest of his physical remains. It is occasionally on display at the *Musée de l'Homme* in Paris.

Copernicus died in 1543 and was buried in an unmarked churchyard grave in Frombork Cathedral in Poland. Archaeologists thought they found the site and the body in 2005 and the identity was confirmed by a DNA match to several hairs discovered in a book owned by the Polish astronomer. The hairs, like Descartes' skull, were not restored to the body. They are kept at Uppsala University but only for forensic interests; they are not seen in public.

A similar thing happened to Tycho Brahe. His body was exhumed in 1901, the 300-year anniversary of his mysterious death. Suspiciously high levels of mercury in the remains encouraged the speculation that he had been poisoned, perhaps by his young colleague Johannes Kepler or by the Danish King Christian IV. Reburied without loss of any bodily parts, he was again bought into daylight in 2010. This time the analysis concluded insufficient amounts of mercury to be the cause of death. It also found a greenish stain in the area of Tycho's nose, with traces of copper and zinc. The prosthetic nose was apparently not of silver and gold as in the legend, but a more proletariat brass. There was, however, evidence of the highlife in the small amount of gold in his hair, consistent with the alchemy of his time that prescribed an elixir of wine flecked with gold to promote longevity. These small samples of Tycho's body were taken for investigative purposes, but none were displayed or revered, and he was reburied more-or-less intact.

And of course there was Einstein's brain, removed and stored in pieces, allegedly in mayonnaise jars. Pierre-Simon Laplace was buried in 1827, also without his brain. Appropriated by his personal physician, it was reported to be smaller than average. It was taken on tour, all in one piece, throughout England.

The dismemberment of Galileo's remains is special, and not just because of his contentious role in making the case that the Earth rotates. Both the details of the event and the subsequent display of the pieces make for an intriguing and amusing story about the relationship between science and society. The vertebra showed up at the University of Padua in 1823 and is still featured in the hall of the faculty of sciences along with the rostrum from which he delivered his lessons—a rough structure with eight steps up to a lecturer's pulpit. One of the fingers, described as the index finger from Galileo's right hand, went immediately into the *Biblioteca Laurenziana* in Florence, where it was placed in a glass urn, much like a crystal reliquary in which you would find preserved bits of saints. It was mounted on a marble pedestal and presented with a Latin inscription, translated,

> This is the finger, belonging to the illustrious hand that ran through the skies, pointing at the immense spaces, and singling out new stars, offering to the senses a marvelous apparatus of crafted glass, and with wise daring they could reach where neither Enceladus nor Tiphaeus could ever reach.

The image of Galileo gesturing to the sky with his right hand and raised index finger features in several artistic renderings of the scientist. A seventeenth-century engraving shows him presenting a telescope to the Muses and directing their attention to an unlikely image of the heliocentric model that hovers overhead. A later painting puts the English writer John Milton in Galileo's Arcetri home, looking through the telescope as his host points out the immense spaces and new stars it reveals. It was a common pose for immortalizing astronomers. Statues of Copernicus usually have him with right hand and index finger raised. The sixteenth-century portrait of Ptolemy has him holding a Jacob's staff, an adjustable and calibrated t-square used to measure the angular separation between stars, in his right hand while pointing to the heavens with the index finger of his left. Plato, not an astronomer but a philosophical hero of Galileo's, stands at center of Raphael's famous *School of Athens* pointing up (right handed) next to Aristotle who gestures with open palm toward the ground.

Galileo's finger in its glass cup and mounted on inscribed marble is now in the History of Science Museum in Florence. But apparently it's not his index finger. A mid-twentieth-century analysis revealed that it had been misidentified. What stands upright in the display is Galileo's middle finger. It seems a defiant gesture in our cultural context, but was it meant to be? And how did the confusion happen, mistaking one finger for another?

The disinterment and defingering of Galileo's body were well documented, with an official notary present to record the names of the participants. It was, apparently, not the pious and respectful restoration of good relations between scientist and church that Viviani had planned. At 6 p.m. on March 12, 1773, with several onlookers, Anton Francesco Gori, a scholar of Etruscan antiquities, borrowed a knife from Giovanni Targioni Tozzetti to remove the fingers and other pieces from the body of the revered scientist. Tozzetti was a physician, with an interest in botany and the history of science. Also participating in the taking of the finger was Antonio Cocchi, a doctor and surgeon. Their own written accounts of the activities list the finger, the one that went into the urn and on the pedestal for display, as the *ditto indice*, the index finger. On their authority it was described this way until the revision by reexamination in the twentieth century. In its withered and bony condition it would take an anatomical expert to identify which of the four non-thumb fingers is in the cup, but at the time of the cutting, when it was attached to the hand and with two trained physicians in the room, it seems unlikely to mistake one finger for another.

Did they do it on purpose, sneak a rude gesture into the celebration of Galileo's physical remains while (most of) the rest of the body was forever hidden away within a cathedral of the church? The upright middle-finger was a recognizable insult at the time. There is reference as early as the fifth century BC. The Greek playwright Aristophanes wrote an affront to Socrates in *The Clouds*

Socrates: Tell me what you know.

Strepsiades: Ever since I was a boy, it's meant this [sticking out his middle finger.]

Socrates: You rustic moron.

The gesture was a popular insult among Romans, calling the middle finger the *digitus impudicus*, the impudent finger. The emperor Caligula forced his enemies and others he wished to demean to kiss his middle finger.

With a cultural context in which the relics of saints and martyrs were sacred and meaningful, and in which the *digitus impudicus* was a recognized rude insult, it is hard to resist at least the speculation that Galileo's raised middle-finger was meant as a defiant poke at the church. He left the church with the parting words, *Eppur si muove,* and with a parting gesture.

Timeline of Important Events and Ideas, Evidence and Theories Regarding the Rotation of the Earth

The evidence that indicates the Earth rotates is in **bold**.

Fourth century BC A Pythagorean model of the cosmos, credited to Philolaus, had a fire burning at the center of the universe while the Earth orbited the fire and rotated on its axis once a day. The reasoning was mostly metaphysical, a reverence for both fire and numbers, but the modeled motions of celestial objects roughly matched what was observed in the sky.

Heraclides eliminated the central fire put the Earth at the center of the universe, rotating once a day. In this model, the stars never move, but the Sun, Moon, and planets slowly orbit the Earth at different rates.

Aristotle applied both physics (the science of physical things on the Earth) and astronomy to compose a cosmology that required a motionless Earth at the center of the universe. The theory was backed by celestial evidence—there is no seasonal shift in the pattern of fixed stars—and the terrestrial evidence that a solid object projected straight up falls straight down. Interpretation of the projectile evidence depended on the Aristotelian understanding of natural motion.

Third century BC Aristarchus proposed a heliocentric model of the universe with the Earth rotating on its axis once a day and revolving around the Sun once a year. There is no record of his reasoning or evidence in support of the idea.

Second century AD Ptolemy assembled revisions to the Aristotelian model, keeping the Earth stationary and located at, or at least near, the center of the universe. He used deferents, epicycles, eccentrics, and equants to match the predictions of the model with astronomical observations, that is, to save the celestial phenomena. He referred to the usual terrestrial evidence such as the atmosphere staying in place and the perpendicular fall of a dropped object to prove the Earth does not rotate.

Fourteen century AD Nicole Oresme used the concept of impetus to argue that no terrestrial phenomena could be evidence one way or the other regarding the rotation of the Earth, since a dropped object would retain its horizontal impetus and follow the ground if the Earth were rotating.

1543 Nicolaus Copernicus published *On the Revolutions of the Heavenly Spheres* with a cosmological model that put the Earth in annual orbit around the Sun while rotating daily on its axis. There was no specific evidence for the rotation, but the full model was credible for its coherence and systematic account of celestial phenomena. Copernicus used Aristotelian principles of physics and astronomy as the basis of the system.

1588 Tycho Brahe proposed a hybrid model of the cosmos in which the Earth is stationary at the center. The Sun orbits the Earth once a day, while the planets orbit the Sun. In part, his motivation was the result of his precise observations of the stars that revealed no annual parallax in their positions.

1600 William Gilbert described the Earth as a great magnet and claimed that **its magnetism causes the Earth to rotate**.

1609 Galileo used a telescope to show the phases of Venus, moons orbiting Jupiter, the rough surface of the Moon, and the movement of Sun spots. These observations contributed to the evidence for the Copernican model of the solar system.

1609 Johannes Kepler revised the heliocentric model to have elliptical orbits, thereby eliminating all of the epicycles, eccentrics, equants, and void points. A magnetic-like force from the rotating Sun, the *vis motrix*, pushes and pulls the planets in their elliptical trajectories. The same force from the Earth moves the Moon. Thus, **the orbit of the Moon is caused by the rotation of the Earth**.

1632 Galileo published the *Dialogue Concerning the Two Chief World Systems* in which he argued that the Copernican model is more likely to be true because it is simpler than the Ptolemaic, and that the tower argument, in which a dropped object to falls straight down, cannot prove one way or the other if the Earth rotates. His evidence for rotation was the **the ocean tides that are caused by the Earth's two motions, daily rotation and annual revolution.**

1633 Galileo was condemned by the Inquisition for advocating the heliocentric model of the solar system in which the Earth moves.

1660's Vincenzo Viviani observed the precession that would later be demonstrated in the Foucault pendulum and celebrated as evidence that the Earth rotates. He dismissed it as a nuisance.

1672 Jean Richer measured the period of a pendulum clock to be longer near the equator than in Paris. Using the Newtonian theory of gravity and dynamics, this was interpreted as showing **the Earth bulges at the equator, caused by its rotation.**

1687 Isaac Newton published *The Mathematical Principles of Natural Philosophy* in which the heliocentric cosmology was explained by the central force of gravity and the dynamics of planetary orbits. The rotating-bucket thought-experiment was presented to show that absolute rotation is detectable.

1727 James Bradley and Samuel Molyneus detected stellar aberration, a change in the apparent direction of incoming starlight caused by the Earth moving as it revolves around the Sun.

1737 An expedition to Lapland measured the on-the-ground distance of one degree of latitude and concluded that the shape of the Earth is an oblate spheroid. **The Earth bulges at the equator, caused by its rotation.** This evidence was interpreted using Newtonian mechanics and its assumption of absolute space and absolute rotation.

1803 Careful measurements in a mineshaft in Schlebusch, Germany showed the **eastward deflection of a falling stone, a Coriolis effect caused by the rotation of the Earth.** This evidence was interpreted using Newtonian mechanics and had been predicted by Newton.

1805 Pierre Laplace, "The rotation of the Earth must be established with certitude that can be provided by the physical sciences. A direct proof of this phenomenon should be of interest to geometers and physicists alike."

1838 Friedrich Bessel measured stellar parallax, a change in the apparent position of a star caused by the Earth being in different positions as it revolves around the Sun.

1851 Léon Foucault built and publically demonstrated a pendulum that precessed slowly clockwise. Using the Newtonian theory of dynamics, the precession of the so-called **Foucault pendulum is a Coriolis effect, caused by the rotation of the Earth.**

1852 Foucault built, named, and demonstrated a gyroscope that precessed with a period of one sidereal day. The **gyroscope maintains its alignment with the fixed stars as the Earth rotates beneath it.**

1883 Ernst Mach published *The Science of Mechanics* and argued that the strict standards of scientific empiricism required rejecting Newton's concept of absolute space as the reference of rotation. Only relative rotation is real, and there is no physical difference between the Ptolemaic and Copernican systems, only a choice of reference frame for the purpose of description and convenience.

1902 A controlled, indoor measurement was made of the **eastward deflection of a falling ball, a Coriolis effect caused by the rotation of the Earth**. The tower used was 23 m high, constructed for this experiment and later used to measure gravitational time-dilation as predicted by the general theory of relativity.

1915 Albert Einstein introduced the general theory of relativity that achieved some of the standards of Mach's principle to make rotation only relative to other physical objects.

1919 The dynamo theory of the Earth's magnetic field was proposed. The **rotation of the Earth is a necessary condition for the sustained magnetic field**.

1925 Albert Michelson used the Sagnac effect of the special theory of relativity to measure **the difference in travel-times for light going in opposite directions around a rectangular path, a difference caused by the rotation of the Earth**.

1990 The Galileo spacecraft on its way to Jupiter sent a video of the Earth. **It showed 25 hours, a full turn of the Earth rotating.**

1992 The Catholic church rehabilitated Galileo, without explicitly saying whether the Earth rotates or not.

Notes

Chapter 1. On Uncertainty

"The crucial thing is being able to move the earth …"

> The comment is from the character Simplicio, advocate of the old world system in the *Dialogue Concerning the Two Chief World Systems*. Galileo (1953, originally published in 1632) p. 122.

"vehemently suspected of heresy,"

> The Papal Condemnation of Galileo, 1633. The source is Santillana (1955) p. 310.

"contrary to the senses and Holy Scripture."

> The Papal Condemnation of Galileo, 1633. The source is Santillana (1955) p. 307.

"In regard to my writing of the Dialogue …"

> Galileo's fourth deposition, June 21, 1633. The source is Finocchiaro (1989) p. 287.

"I ought no less …"

> Descartes (1993, originally published in 1641) p. 11.

"Like you, I accepted the Copernican …"

> Galileo letter to Kepler, 1597. The source is Santillana (1955) p. 11.

"Be of good cheer, Galileo …"

> Kepler letter to Galileo, 1597. The source is Santillana (1955) p. 15.

Chapter 2. To Save the Phenomena

"trying by violence to bring the appearances into line …"

> The comment is in reference to Pythagorean models of the cosmos that put the Earth in motion. Aristotle (1939, original ca. 350 BC) p. 217.

"Heraclides supposed that …"

> Simplicius, sixth century AD, *Commentary on Aristotle's On the Heavens*, cited in Lloyd, G. (1970) p. 95.

Chapter 3. Aristotle's Standard Model

"To be a good investigator …"

> Aristotle (1939, original ca. 350 BC) p. 227.

"To be ignorant of motion is to be ignorant of nature."

> This is a slogan attributed to Aristotle and Peripatetics as a group. It is drawn from the more prosaic, "Since nature is the principle of movement and change, and it is Nature that we are studying, we must understand what 'movement' is; for, if we do not know this neither do we understand what Nature is."
>
> Aristotle (1957, original ca. 350 BC) p. 191.

"What are the uniform and ordered movements …"

> The question is Plato's but the wording is attributed to Sosigenes, a second-century AD commentator and reported by Simplicius. The modern source is Lloyd (1970) p. 84.

"If seven ants were to be placed on a potter's wheel …"

> Vitruvius (1999, original ca. 30 BC) page 111.

"It so happens that the earth and the Universe …"
    Aristotle (1939, original ca. 350 BC) p. 245.

"If then any particular portion is incapable …"
    Aristotle (1939, original ca. 350 BC) p. 247.

"the order of the world is eternal,"
    Aristotle (1939, original ca. 350 BC) p. 243.

"there would have to be passings and turnings …"
    Aristotle (1939, original ca. 350 BC) p. 243.

"whether [the Earth] move …"
    Aristotle (1939, original ca. 350 BC) p. 243.

"heavy objects, if thrown forcibly upward …"
    Aristotle (1939, original ca. 350 BC) p. 245.

"from these considerations …"
    Aristotle (1939, original ca. 350 BC) p. 245.

Chapter 4. Tinkering with the Standard Model

"Certain people …"
    Ptolemy (1984, original second century AD) p. 44.

"one of the greatest …"
    NASA web site, https://apod.nasa.gov/apod/ap971108.html. Accessed on January 7, 2019.

"[Aristarchus'] hypotheses are …"
    Archimedes, *Sand Reckoner*, cited in Barbour (2001) p. 188.

"In a *yuga*, the revolutions …"
    Aryabhata (1930, original early sixth century AD) p. 9.

"I have seen the astrolabe …"

Al-Biruni, quoted in Nasr, S. (1993) p. 135.

"For it is the same …"

Al-Biruni, quoted in Nasr, S. (1993) p. 136.

"It is not possible to attribute primary motion …"

Tusi, quoted in Ragep, F. (2001) p. 147.

"a principle of rectilinear inclination …"

Tusi, quoted in Ragep, F. (2001) p. 147.

"the uniform and ordered movements …"

The question is Plato's but the wording is attributed to Sosi-genes, a second-century AD commentator and reported by Simplicius. The modern source is Lloyd (1970) p. 84.

"Now let no one, considering the complicated nature …"

Ptolemy (1984, original second century AD) p. 600.

"The shell game that we play …"

Feynman (1985) p. 128.

"Certain people …"

Ptolemy (1984, original second century AD) p. 44.

"such a notion is quite ridiculous."

Ptolemy (1984, original second century AD) p. 45.

"there is perhaps nothing in celestial phenomena …"

Ptolemy (1984, original second century AD) p. 45.

"the revolving motion of the earth …"

Ptolemy (1984, original second century AD) p. 45.

"if they said that the air is carried around …"

Ptolemy (1984, original second century AD) p. 45.

"if those objects too were carried around …"

Ptolemy (1984, original second century AD) p. 45.

"[The projector] impresses a certain …"

Buridan, *Quaestions super octo libros physicorum*, cited in Kuhn (1957) p. 119.

"… an arrow shot straight into the air …"

Oresme, *Le livre du ciel et du monde*, cited in Kuhn (1957) p. 116.

"I suppose that local motion …"

Oresme, *Le livre du ciel et du monde*, cited in Kuhn (1957) p. 115.

Chapter 5. Moving the Earth

"… to ascribe movement to the Earth …"

This is in Copernicus' Preface to *On the Revolutions of the Heavenly Spheres*, a letter directed to Pope Paul III. The source is Kuhn (1957) p. 137.

"a relation that nature abhors"

Georg Joachim Rheticus, quoted in Hoskin, M. and Gingerich, O. (1999) p. 87.

"a second Ptolemy."

Tycho Brahe, quoted in Westman (1975) p. 307.

"What are the uniform and ordered movements …"

The question is Plato's but the wording is attributed to Sosigenes, a second-century AD commentator and reported by Simplicius. The modern source is Lloyd (1970) p. 84.

"for it is not necessary that these hypotheses …"

Andreas Osiander, the unauthorized and anonymous Introduction to *On the Revolutions of the Heavenly Spheres*, Copernicus (2002, originally published in 1543) p. 1.

"written by a jackass for the use of other jackasses"
>   Johannes Kepler, quoted in Santillana (1955) p. 101.

"ignorant and presumptuous ass"
>   Giordano Bruno quoted in Polanyi (1962) p. 155.

"And since it is the heavens which contain …"
>   Copernicus (2002, original 1543) p. 13.

"because the neighboring air …"
>   Copernicus (2002, original 1543) p. 18.

"we must confess that in comparison …"
>   Copernicus (2002, original 1543) p. 18.

"And things are as when Aeneas said in Virgil …"
>   Copernicus (2002, original 1543) p. 17.

"You see therefore that for all these reasons …"
>   Copernicus (2002, original 1543) p. 19.

Chapter 6. The Best of Both Worlds

"… the earth, that hulking, lazy body, unfit for motion …"
>   Tycho Brahe, quoted in Gingerich, O., and Voelkel, J. (1998) p. 24.

"this brave o'erhanging firmament …"
>   William Shakespeare (1601) *Hamlet*, Act 2, scene 2.

"yond same star …"
>   William Shakespeare (1601) *Hamlet*, Act 1, scene 1.

"a second Ptolemy"
>   Tycho Brahe, quoted in Westman (1975) p. 307.

"Copernicus nowhere offends …"
>   Tycho Brahe, quoted in Gingerich (1993) p. 33.

"Can we formulate physical laws …"

> Einstein and Infeld (1938) pp. 224–225.

Chapter 7. On Skepticism

"[They] seem to me to be making the mistake of judging …"

> Ptolemy (1984, original second century AD) p. 44.

"… idle babblers, ignorant of mathematics, …"

> Copernicus (2002, original 1543) p. 2. This is in the author's Preface, directed to Pope Paul III.

"Observation contaminated by …"

> Scheffler (1982) p. 14.

"The single most striking feature …"

> Kuhn (1977) p. 228.

"the death of expertise"

> Nichols (2017).

Chapter 8. The Two Chief World Systems

"And yet it moves."

> Galileo, allegedly muttered to himself just after his conviction by the Inquisition in 1633.

"Philosophy is written in this grand book …"

> Galileo (1957, originally published in 1623) pp. 237–238.

"The Explication of the Three-Fold Motion …,"

> Galileo's copy of Copernicus' *On the Revolutions of the Heavenly Spheres*, quoted and displayed in Gingerich (2004) p. 145.

"Ptolemy introduces vast epicycles …"

> Galileo (1953, originally published in 1632) p. 342.

"whatever motion comes to be attributed to the earth …"
    Galileo (1953, originally published in 1632) p. 114.

"The true method …"
    Galileo (1953, originally published in 1632) pp. 114–116.

"Rather, we should not judge 'simplicity' …"
    Ptolemy (1984, original second century AD) p. 600.

"I tested the instrument of Galileo's …"
    Martin Horky in 1610, quoted in Feyerabend (1975) p. 123.

"Aristotle says then that the most certain proof …"
    Galileo (1953, originally published in 1632) p. 139.

"For to expect the rock …"
    Galileo (1953, originally published in 1632) pp. 140–141.

"falls to the foot of the mast …"
    Galileo (1953, originally published in 1632) p. 141.

"Our discourse must relate to …"
    Galileo (1953, originally published in 1632) p. 113.

"Without experiment …"
    Galileo (1953, originally published in 1632) p. 145.

"if the terrestrial globe were immoveable …"
    Galileo (1953, originally published in 1632) p. 417.

"when we confer upon the globe …"
    Galileo (1953, originally published in 1632) p. 417.

"whatever motion comes to be attributed to the earth …"
    Galileo (1953, originally published in 1632) p. 114.

Chapter 9. New Astronomy and the Great Magnet

"Hence the entire terrestrial globe ..."

Gilbert (1958, original 1600) p. 326.

"The method of investigation ..."

Galileo (1953, original 1632) p. 417.

"A mathematical point ..."

Kepler (1992, original 1609) p. 54.

"Copernicus, restorer of astronomy."

Gilbert (1958, original 1600) p. 358.

"Surely that is superstition ..."

Gilbert (1958, original 1600) pp. 321–322.

"... all magnetic bodies ..."

Gilbert (1958, original 1600) p. 335.

"From these arguments ..."

Gilbert (1958, original 1600) p. 327.

"all the circumfused effluences ..."

Gilbert (1958, original 1600) p. 340.

"But these are old-wives' imaginings ..."

Gilbert (1958, original 1600) p. 337.

"The goal of magnetic virtue ..."

Jacques Grandami (1648) *Nova demonstration immobilitatis terrae petita ex virtute*, quoted in Baldwin (1985) p. 168.

"the orb by an arcane force ..."

Sylvestre di Pietra-Sancta (1634) *De simbolis heroicis*, quoted in Baldwin (1985) p. 162.

Chapter 10. Rotational Dynamics and Absolute Space

"I will communicate to you a fancy of my own …"

> Isaac Newton (1679) letter to Robert Hooke, quoted in Ball (1893) p. 142.

"there are circumstances in which mathematics …"

> Feynman (1963) p. 20–26.

"First, That all Coelestial Bodies whatsoever …"

> Robert Hooke (1674) *Attempt to Prove the Motion of the Earth*, quoted in Hoskin (1997) p. 148.

"That these attractive powers …"

> Robert Hooke (1674) *Attempt to Prove the Motion of the Earth*, quoted in Hoskin (1997) p. 149.

"That all bodies whatsoever …"

> Robert Hooke (1674) *Attempt to Prove the Motion of the Earth*, quoted in Hoskin (1997) pp. 148–149.

"… if our earth were not a little …"

> Newton (1995, originally published in 1687) p. 341.

"Vous avez confirmé …"

> Voltaire (1737) letter to Maupertuis, quoted in Poincaré (2003, original 1914) p. 275.

"… will not descend the perpendicular, …"

> Isaac Newton (1679) letter to Robert Hooke, quoted in Ball (1893) p. 143.

"… and that very considerably, …"

> Robert Hooke (1680) letter to Isaac Newton, quoted in Ball (1893) p. 148.

"And far from failing to follow …"

> Galileo (1953, originally published in 1632) p. 233.

Chapter 11. Foucault's Pendulum

"You are invited to come see the Earth turn, …"

Léon Foucault (February 2, 1851) printed invitation, quoted in Tobin (2003) p. 141.

"Although the rotation of the earth …"

Laplace (1839, originally published in 1805) p. 573.

"There have been made, in Italy and Germany, …"

Laplace (1839, originally published in 1805) p. 590.

"Gentlemen of the Astronomical Society, …"

Herschel (1841) p. 453.

"When all the precautions described …"

Shapiro (1962) p. 1081.

"Wednesday, January 8, 2 a.m.: the pendulum turned …"

Léon Foucault (1851) journal entry, quoted in Tobin (2003) p. 139.

Chapter 12. Mach's Principle

"These two propositions, 'the earth turns round,' and …"

Poincaré (1952, originally published in 1902) p. 117.

"in its own nature, without …"

Newton (1995, originally published in 1687) p. 13.

"The Newtonian theory of gravitation …"

Mach (1911, originally published in 1909) p. 56.

"No one is competent to say …"

Mach (1960, originally published in 1883) p. 284.

"the pendulum turned in the direction …'

Léon Foucault (1851), quoted in Tobin (2003) p. 139.

"The motions of the universe …"

> Mach, E. (1960, originally published in 1883) p. 284.

"arbitrary fictions of our imagination."

> Mach (1960, originally published in 1883) p. 284.

"a catalogue and not a system."

> Albert Einstein in a published discussion among several scientists on the theory of relativity, in Leclerc, M. ed. (1922) p. 112.

Chapter 13. Relativity

"Rotation is thus relative in Einstein's theory."

> de Sitter (1917) p. 532.

"… besides observable objects, another thing, …"

> Einstein (1920) a Lecture Presented on 5th May, 1920 in the University of Leiden, printed in Einstein (1922) p. 17.

"For it is the same …"

> Al-Biruni, quoted in Nasr, S. (1993) p. 136.

"I suppose that local motion …"

> Oresme, *Le livre du ciel et du monde*, cited in Kuhn (1957) p. 115.

"… one had to cram all this stuff into one's head …"

> Einstein (1969) pp. 17–19.

"… it seems probable that most of the grand underlying principles …"

> Michelson (1894) p. 159.

"Thus, in our theory …"

> Sciama (1953) p. 41.

"Hence, the condition for …"

> Braeck, Gron, & Farup (2017) p. 13.

"All of the centrifugal and Coriolis effects …"
 Braeck, Gron, & Farup (2017) p. 13.

Chapter 14. The Final Frontier

"Don't think; but look!"
 Wittgenstein (1953) p. 31.

"You can observe a lot just by watching."
 Berra (2008)

"the gravomagnetic field …"
 Sciama (1953) p. 41.

"making the mistake …"
 Ptolemy (1984, original second century AD) p. 44.

Chapter 15. All Things Considered

"The truth for which Galileo …"
 Poincaré (19907, originally published in 1905) p. 141.

"taking into account …"
 Ptolemy (1984, original second century AD) p. 44.

"the printing and publication …"
 Catholic College of Cardinals (1822), quoted in Bartolotta (2017) p. 85.

"The Bible does not concern itself …"
 Pope John Paul II (1992) paragraph 12.

"He rejected the suggestion …"
 Pope John Paul II (1992) paragraph 5.

"In regard to my writing of the Dialogue …"
 Galileo's fourth deposition, June 21, 1633, quoted in Finocchiaro (1989) p. 287.

"not to defend it …"

> The Papal Condemnation of Galileo, 1633, quoted in Santillana (1955) p. 307.

"If there were a real proof that the Sun …"

> Bellarmine (1615) *Letter to Foscarini*, quoted in Koestler 1959, pp. 447–448.

"Socrates: Tell me what you know …"

> Aristophanes (2002, original 423 BC), p. 99.

Bibliography

Aristophanes (2002, original 450 BC) *Lysistrata and Other Plays*, translated by A. Sommerstein (Penguin, London).

Aristotle (1939, original ca. 350 BC) *On the Heavens*, translated by W. Guthrie (Loeb Classical Library, Cambridge, USA).

Aristotle (1957, original ca. 350 BC) *Physics*, translated by P. Wicksteed and F. Cornford (Loeb Classical Library, Harvard University Press, Cambridge, USA).

Aryabhata (1930, original early sixth century AD) *The Aryabhatia of Aryabhata*, translated by W. Clark (University of Chicago Press, Chicago).

Baldwin, M. (1985) "Magnetism and the Anti-Copernican Polemic," *Journal of the History of Astronomy*, **16**, 155-174.

Ball, W. (1893) *An Essay on Newton's "Principia"* (MacMillan & Co., New York).

Barbour, J. (2001) *The Discovery of Dynamics* (Oxford University Press, New York).

Barbour, J. and Pfister, H., eds. (1995) *Mach's Principle: From Newton's Bucket to Quantum Gravity* (Birkhauser, Boston).

Bartolotta, K. (2017) *The Inquisition: The Quest for Absolute Religious Power* (Lucent Press, New York).

Berra, Y. (2008) *You Can Observe a Lot Just by Watching: What I've Learned About Teamwork from the Yankees and Life* (John Wiley & Sons, Hoboken).

Braeck, S., Gron, O., and Farup, I. (2017) "The Cosmic Causal Mass," *Universe*, **3** (38), 1-21.

Copernicus, N. (2002, originally published in 1543) *On the Revolutions of the Heavenly Spheres*, translated by C. Wallace (Running Press, Philadelphia).

de Sitter, W. (1917) "On the relativity of rotation in Einstein's theory," *Royal Netherlands Academy of Arts and Sciences Proceedings*, **191**, 527-532.

Descartes, R. (1993, originally published in 1641) *Meditation on First Philosophy*, translated by D. Cress (Hackett Publishing, Indianapolis).

Einstein, A. (1922) *Sidelights on Relativity*, translated by G. Jeffery and W. Perrett (Methuen & Co., London).

Einstein, A. (1969) "Autobiographical Notes," in *Albert Einstein: Philosopher and Scientist*, P. ed. Schlipp, 3rd edition (Open Court, LaSalle) pp. 1-95.

Einstein, A. and Infeld, L. (1938) *The Evolution of Physics: The Growth of Ideas from Early Concepts to Relativity and Quanta* (Simon and Schuster, New York).

Feyerabend, P. (1975) *Against Method* (Verso, London).

Feynman, R. (1963) *Lectures on Physics*, volume 1 (Addison-Wesley, Reading, USA).

Feynman, R. (1985) *QED: The Strange Theory of Light and Matter* (Princeton University Press, Princeton).

Finocchiaro, M. (1989) *The Galileo Affair: A Documentary History* (University of California Press, Berkeley).

Galileo (1953, originally published in 1632) *Dialogue Concerning the Two Chief World Systems*, translated by S. Drake (University of California Press, Berkeley).

Galileo (1957, originally published in 1623) *The Assayer*, translated by S. Drake (Doubleday & Company, New York).

Gilbert, W. (1958, original 1600) *De Magnete*, translated by P. Mottelay (Dover Publications, Mineola).

Gingerich, O. (1993) *The Eye of Heaven* (American Institute of Physics, New York).

Gingerich, O. (2004) *The Book Nobody Read* (Walker Publishing, New York).

Gingerich, O. and Voelkel, J. (1998) "Tycho Brahe's Copernican Campaign," *Journal for the History of Astronomy*, **29** (part 1), 1-34.

Herschel, W. (1841) "An Address Delivered at the Annual General Meeting of the Royal Astronomical Society, February 12, 1841, Presenting the Honorary Medal to M. Bessel," *Royale Astronomical Society Transactions*, **xii**, 442-453.

Hoskin, M. (1997a) "Newton and Newtonianism," in *The Cambridge Illustrated History of Astronomy* (Cambridge University Press, Cambridge, UK) pp.144-197.

Hoskin, M., ed. (1997b) *The Cambridge Illustrated History of Astronomy* (Cambridge University Press, Cambridge, UK).

Hoskin, M. (1999) *The Cambridge Concise History of Astronomy* (Cambridge University Press, Cambridge, UK).

Hoskin, M. and Gingerich, O. (1999) "Medieval Latin Astronomy" in M. Hoskin, ed. *The Cambridge Concise History of Astronomy* (Cambridge University Press, Cambridge, UK) pp. 68-87.

John Paul II (1992) "Allocution," *L'Osservatore Romano*, number 44, November 4.

Kepler, J. (1992, originally published in 1609) *New Astronomy*, translated by W. Donahue (Cambridge University Press, Cambridge, UK).

Koestler, A. (1959) *The Sleepwalkers: A History of Man's Changing Vision of the Universe* (Hutchinson, London).

Kuhn, T. (1957) *The Copernican Revolution* (Harvard University Press, Cambridge, USA).

Kuhn, T. (1977) *The Essential Tension* (University of Chicago Press, Chicago).

Laplace, P. (1839, originally published in 1805) *Mécanique Céleste*, volume 4, translated by N. Bowditch (Little & Brown, Boston).

Leclerc, M. (1922) "Séance du 6 avril 1922: La Théorie de la Relativité," *Bulletin de la Société Française de Phiosophie*, **22**, 91-112.

Lloyd, G.E.R. (1970) *Early Greek Science: Thales to Aristotle* (W.W. Norton, New York).

Mach, E. (1911, originally published in 1909) *History and Root of the Principle of the Conservation of Energy*, translated by P. Jourdain (Open Court, London).

Mach, E. (1960, originally published in 1883) *The Science of Mechanics*, 6th edition, translated by T. McCormack (Open Court, Chicago).

Michelson, A. (1894) "Dedication of Ryerson Physical Laboratory," reprinted in the University of Chicago *Annual Register*, July 1895-July 1896, p. 159.

Nasr, S. (1993) *An Introduction to Islamic Cosmological Doctrines* (SUNY Press, Albany).

Newton, I. (1995, originally published in 1687) *The Mathematical Principles of Natural Philosophy*, translated by A. Motte (Prometheus Books, Amherst).

Nichols, T. (2017) *The Death of Expertise: The Campaign Against Established Knowledge and Why It Matters* (Oxford University Press, New York).

Poincaré, H. (1907, originally published in 1905) *The Value of Science*, translated by G. Halsted (The Science Press, New York).

Poincaré, H. (1952, originally published in 1902) *Science and Hypothesis*, translated by W. Greenstreet (Dover, New York).

Poincaré, H. (2003, originally published in 1914) *Science and Method*, translated by F. Maitland (Dover, New York).

Polanyi, M. (1962) *Personal Knowledge* (Routledge, London).

Ptolemy (1984, original second century AD) *Almagest*, translated by G. Toomer (Duckworth, London).

Ragep, F. (2001) "Tusi and Copernicus: The Earth's Motion in Context," *Science in Context*, **14**, 145-163.

Reichenbach, H. (1958) *The Philosophy of Space and Time*, translated by M. Reichenbach and J. Freund (Dover, New York).

Rothman, T. (2017) "The Forgotten Mystery of Inertia," *American Scientist*, **105** (6), 344-355.

Santillana, G. (1955) *The Crime of Galileo* (University of Chicago Press, Chicago).

Schapiro, A. (1962) "Bath-Tub Vortex," *Nature* **196**, 1080-1081.

Scheffler, I. (1982) *Science and Subjectivity* (Hackett Publishing Company, Indianapolis).

Sciama, D. (1953) "On the Origin of Inertia," *Monthly Notices of the Royal Astronomical Society*, **113**, 34-42.

Staley, R. (2013) "Ernst Mach on Bodies and Buckets," *Physics Today*, December 2013, pp. 42-47.

Tobin, W. (2003) *The Life and Science of Léon Foucault* (Cambridge University Press, Cambridge UK).

Vitruvius (1999, original ca. 30 BC) *Ten Books on Architecture*, translated by I. Rowland (Cambridge University Press, Cambridge, UK).

Westman, R. (1975a) "Three Responses to Copernican Theory: Johannes Praetorius, Tycho Brahe, and Michael Maestlin," in *The Copernican Achievement* (University of California Press, Berkeley) pp. 285-345.

Westman, R., ed. (1975b) *The Copernican Achievement* (University of California Press, Berkeley).

Wittgenstein, L. (1953) *Philosophical Investigations*, translated by G. Anscombe (Blackwell, Oxford).

Index

Printed in the United States
By Bookmasters